T0133972

APPLIED PHARMACEUTICAL SCIENCE AND MICROBIOLOGY

Novel Green Chemistry Methods
and Natural Products

APPLIED PHARMACEUTICAL SCIENCE AND MICROBIOLOGY

Novel Green Chemistry Methods
and Natural Products

Edited by
Debarshi Kar Mahapatra, PhD
Swati Gokul Talele, PhD, MPharm
A. K. Haghi, PhD

First edition published [2021]

Apple Academic Press Inc.
1265 Goldenrod Circle, NE,
Palm Bay, FL 32905 USA

4164 Lakeshore Road, Burlington,
ON, L7L 1A4 Canada

CRC Press
6000 Broken Sound Parkway NW,
Suite 300, Boca Raton, FL 33487-2742 USA

2 Park Square, Milton Park,
Abingdon, Oxon, OX14 4RN UK

Library and Archives Canada Cataloguing in Publication

Title: Applied pharmaceutical science and microbiology : novel green chemistry methods and natural products / edited by Debarshi Kar Mahapatra, PhD, Swati Gokul Talele, PhD, MPharm, A. K. Haghi, PhD.

Names: Mahapatra, Debarshi Kar, editor. | Talele, Swati Gokul, editor. | Haghi, A. K., editor.

Description: Includes bibliographical references and index.

Identifiers: Canadiana (print) 20200296833 | Canadiana (ebook) 20200297023 | ISBN 9781771888912 (hardcover) | ISBN 9781003019565 (ebook)

Subjects: LCSH: Drug development. | LCSH: Pharmaceutical microbiology. | LCSH: Natural products. | LCSH: Green chemistry.

Classification: LCC RM301.25 .A67 2021 | DDC 615.1/9—dc23

Library of Congress Cataloging-in-Publication Data

Names: Mahapatra, Debarshi Kar, editor. | Talele, Swati Gokul, editor. | Haghi, A. K., editor.

Title: Applied pharmaceutical science and microbiology : novel green chemistry methods and natural products / edited by Debarshi Kar Mahapatra, Swati Gokul Talele, A. K. Haghi.

Description: Burlington ON ; Palm Bay, Florida : Apple Academic Press, [2021] | Includes bibliographical references and index. | Summary: "This volume on applied pharmaceutical science and microbiology looks at the latest research on the applications of natural products for drug uses. It focuses on understanding how to apply the principles of novel green chemistry methods in the vital area of pharmaceuticals and covers the important aspects of green microbial technology in the pharmaceutical industry. Chapters include studies on the applications used of natural products in folk and regional medicines, such as for digestive problems, dermatological infections, respiratory diseases, vessel diseases, diarrhea and dysentery, ringworms, boils, fevers (antipyretic), skin and blood diseases, mouth sores, channel discharges, and even cancer. It looks at medical benefit of microbial fermentation for the conservation of nutrients. Other chapters present information on AgNPs fabricated from natural plant and organism extracts for the treatment of cancer and nanoparticles-based drug delivery system as promising technology in microbial pharmaceutics and extensive applications. Other topics cover the structures of phytochemicals biological activities, microbial transformations of natural compounds in the advancement of steroid drugs, spherulites used as a carrier for drug delivery, and much more. Applied Pharmaceutical Science and Microbiology: Novel Green Chemistry Methods and Natural Products provides an abundance of new research on this increasingly important area"-- Provided by publisher.

Identifiers: LCCN 2020033270 (print) | LCCN 2020033271 (ebook) | ISBN 9781771888912 (hardcover) | ISBN 9781003019565 (ebook)

Subjects: MESH: Technology, Pharmaceutical | Microbiological Phenomena | Green Chemistry Technology | Biological Products

Classification: LCC RM301.25 (print) | LCC RM301.25 (ebook) | NLM QV 778 | DDC 615.1/9--dc23

LC record available at https://lccn.loc.gov/2020033270
LC ebook record available at https://lccn.loc.gov/2020033270

ISBN: 978-1-77188-891-2 (hbk)
ISBN: 987-1-00301-956-5 (ebk)

About the Editors

Debarshi Kar Mahapatra, PhD
Assistant Professor, Department of Pharmaceutical Chemistry,
Dadasaheb Balpande College of Pharmacy, Rashtrasant Tukadoji
Maharaj Nagpur University, Nagpur, Maharashtra, India

Debarshi Kar Mahapatra, PhD, is currently working as an Assistant Professor in the Department of Pharmaceutical Chemistry at Dadasaheb Balpande College of Pharmacy, Rashtrasant Tukadoji Maharaj Nagpur University, Nagpur, Maharashtra, India. He was formerly working as an Assistant Professor in the Department of Pharmaceutical Chemistry, Kamla Nehru College of Pharmacy; RTM Nagpur University, Nagpur, India. He has taught medicinal and computational chemistry at both the undergraduate and postgraduate levels and has mentored students in their various research projects. His area of interest includes computer-assisted rational designing and synthesis of low molecular weight ligands against druggable targets, drug delivery systems, and optimization of unconventional formulations. He has published research, book chapters, reviews, and case studies in various reputed journals and has presented his work at several international platforms, for which he has received several awards by a number of bodies. He has also authored the book titled *Drug Design*. Presently, he is serving as a reviewer and editorial board member for several journals of international repute. He is a member of a number of professional and scientific societies, such as the International Society for Infectious Diseases (ISID), the International Science Congress Association (ISCA), and ISEI.

Swati Gokul Talele, PhD
Assistant Professor, Department of Pharmaceutics,
Sandip Institute of Pharmaceutical Sciences,
Savitribai Phule Pune University, Pune, Maharashtra, India

Swati Gokul Talele, PhD, is currently serving as an Assistant Professor, Department of Pharmaceutics, at Sandip Institute of Pharmaceutical

Sciences. She has 18 years of experience in research along with teaching. She has published more than 21 research papers in various reputed international and national journals. She has supervised many MPharm students and currently associated with many research projects.

She has published more than 30 review papers and presented research work at several conferences and received awards by universities. She has also authored the book titled Natural excipients and many chapters and books are in progress. Her interests are in the field of nanotechnology, natural polymers, herbal formulations, and radiolabeling based biodistribution studies.

She is a university-approved and postgraduate teacher of Pharmaceutics. She also worked as a College Examination Officer (CEO) for more than 3 years and a member of the university examination committee. She has delivered an interactive talk at a continuous education program for registered pharmacists. She is a life member of the association of pharmacy teachers of India (APTI) and a member of the Indian Pharmaceutical Congress Association (IPCA).

A. K. Haghi, PhD
Professor Emeritus of Engineering Sciences, Former Editor-in-Chief, International Journal of Chemoinformatics and Chemical Engineering and Polymers Research Journal; Member, Canadian Research and Development Center of Sciences and Culture

A. K. Haghi, PhD, is the author and editor of 165 books, as well as 1000 published papers in various journals and conference proceedings. Dr. Haghi has received several grants, consulted for a number of major corporations, and is a frequent speaker to national and international audiences. Since 1983, he served as a professor at several universities. He is former Editor-in-Chief of the *International Journal of Chemoinformatics and Chemical Engineering* and *Polymers Research Journal* and is on the editorial boards of many international journals. He is also a member of the Canadian Research and Development Center of Sciences and Cultures (CRDCSC), Montreal, Quebec, Canada.

Contents

Contributors

Abutahir
Raghukul College of Pharmacy, Sarvar, Bhopal, Madhya Pradesh, India

Eknath D. Ahire
Department of Pharmaceutics, SSS's, Divine College of Pharmacy, Nampur Road, Satana, Maharashtra, India; Sandip Institute of Pharmaceutical Sciences, Nashik, Maharashtra, India

Niyaz Ahmad
College of Clinical Pharmacy, Imam Abdulrahman Bin Faisal University, Dammam, Saudi Arabia

Wasim Ahmad
Department of Pharmacy, Mohammad Al-Mana College for Medical Sciences, Dammam, Saudi Arabia

Abuzer Ali
College of Pharmacy, Taif University, Haweiah, Taif, Saudi Arabia, Tel.: +966599131866, E-mail: abuzerali007@gmail.com

Amena Ali
College of Pharmacy, Taif University, Haweiah, Taif, Saudi Arabia

Mohd. Amir
College of Clinical Pharmacy, Imam Abdulrahman Bin Faisal University, Dammam, Saudi Arabia

M. R. Badwar
Department of Quality Assurance, Sandip Institute of Pharmaceutical Sciences, Mahiravani, Nashik, Maharashtra, India

Akshada Atul Bakliwal
Department of Pharmaceutics, Sandip Institute of Pharmaceutical Sciences, Mahiravani, Nashik, Maharashtra, India

Sanjay Kumar Bharti
Institute of Pharmaceutical Sciences, Guru Ghasidas Vishwavidyalaya (A Central University), Bilaspur – 495006, Chhattisgarh, India

Gloria Castellano
Department Experimental Sciences and Mathematics, Catholoc University of Valencia "San Vincent Martir," Valencia, Spain

Arijit Chaudhuri
U.S. Ostwal Institute of Pharmacy, Mangalwad – 312024, Rajasthan, India

Sunil Kumar Chauhan
Manav Bharti University, Solan – 173229, Himachal Pradesh, India

Manik Das
Department of Pharmacy, Tripura University (A Central University), Suryamaninagar – 799022, Tripura, India; Department of Pharmaceutical Chemistry, Srikrupa Institute of Pharmaceutical Sciences, Hyderabad – 502277, Telangana, India

Anil G. Jadhav
Department of Pharmacognosy, Sandip Institute of Pharmaceutical Sciences, Mahiravani, Nashik, Maharashtra, India

Souranava Jana
Institute of Pharmaceutical Sciences, Guru Ghasidas Viswavidyalaya (A Central University),
Bilaspur – 495009, Chhattisgarh, India

A. K. Haghi
Professor Emeritus of Engineering Sciences, Former Editor-in-Chief, *International Journal of
Chemoinformatics and Chemical Engineering and Polymers Research Journal*; Member,
Canadian Research and Development Center of Sciences and Culture
E-mail: akhaghi@yahoo.com

Nayana Kale
Sandip Institute of Pharmaceutical Sciences, Nashik, Maharashtra, India

Debarshi Kar Mahapatra
Department of Pharmaceutical Chemistry, Dadasaheb Balpande College of Pharmacy,
Nagpur – 440037, Maharashtra, India, E-mail: mahapatradebarshi@gmail.com

Kuntal Manna
Department of Pharmacy, Tripura University (A Central University), Suryamaninagar – 799022,
Tripura, India

Souvik Mukherjee
Institute of Pharmaceutical Sciences, Guru Ghasidas Viswavidyalaya (A Central University),
Bilaspur – 495009, Chhattisgarh, India, E-mail: mukherjees388@gmail.com

Chandrashekhar D. Patil
Sandip Institute of Pharmaceutical Sciences, Nashik, Maharashtra, India

Vivek Kumar Raman
Manav Bharti University, Solan – 173229, Himachal Pradesh, India,
E-mail: vivekraman72@gmail.com

Heena S. Shah
Department of Pharmaceutics, SSS's, Divine College of Pharmacy, Nampur Road, Satana,
Maharashtra, India; Sandip Institute of Pharmaceutical Sciences, Nashik, Maharashtra, India

Ganesh B. Sonwane
Sandip Institute of Pharmaceutical Sciences, Nashik, Maharashtra, India

Khemchand R. Surana
Sandip Institute of Pharmaceutical Sciences, Nashik, Maharashtra, India

Deepal J. Taank
Sandip Institute of Pharmaceutical Sciences, Nashik, Maharashtra, India

Swati Gokul Talele
Department of Pharmaceutics, Sandip Institute of Pharmaceutical Sciences, Mahiravani,
Nashik, Maharashtra, India, E-mail: swatitalele77@gmail.com

Roshan Telrandhe
Department of Pharmaceutics, Smt. Kishoritai Bhoyar College of Pharmacy, New Kamptee,
Nagpur – 441002, Maharashtra, India

Francisco Torrens
Department of Molecular Chemistry, University of Valencia, Valencia, Spain,
E-mail: torrens@uv.es

Musarrat Husain Warsi
College of Pharmacy, Taif University, Haweiah, Taif, Saudi Arabia

Abbreviations

5-HETE	5-hydroxyeicosatetraenoic acid
5-LO	5-lipoxygenase
ABA	acetyl boswellic acids
ACE	angiotensin 1-converting enzyme
AD	Alzheimer disease
AESG	alcoholic extract of Salai-guggal
AFB1	aflatoxin B1
AFM	atomic force research
AgNPs	silver nanoparticles
AKBA	acetyl-11-keto-β-boswellic acid
ALA	alpha-linolenic acid
AOAs	AO activities
AOs	antioxidants
BAL	bronchoalveolar lavage
BAs	biogenic amines
BAs	boswellic acids
BDNF	brain imitative neurotrophic factor
BHT	butylated hydroxytoluene
CAGR	compound yearly development rate
CaP	carcinoma of the prostate
CAqE	aqueous extract
CAT	catalase
CDKN-1A	cyclin-dependent kinase inhibitor 1A
COX	cyclooxygenase
COX-1	cyclooxygenase-1
CRC	cancer cells
CVDs	cardiovascular diseases
DAPI	4,6-diamidino-2-phenylindole
DHA	docosahexaenoic acid
DKK-1	dickkopf-related protein 1
DM	diabetes mellitus
DTQ	dithymoquinone
EAE	experimental allergic encephalitis

EFSA	European Food Safety Authority
EMA	European Medicines Agency
EO	essential oil
EPA	eicosapentaenoic acid
ERK1/2	extracellular signal-regulated protein kinases 1 and 2
FDA	Food and Drug Administration
FOS	fructooligosaccharides
FOSHU	food of specified health use
FTIR	Fourier transform infrared
GFN	growth front nucleation
GI tract	gastrointestinal tract
GR	glutathione reductase
GRAS	generally regarded as sheltered
GST	glutathione transferase
GTPx	glutathione peroxidize
H_2O_2	hydrogen peroxide
HDL-C	high-density lipoprotein-cholesterol
HLE	human corpuscle enzyme
HMG-CoA	β-hydroxy β-methylglutaryl-CoA
HPLC	high-performance liquid chromatography
HSP-90	heat shock protein 90
hTERT	human telomerase reverse transcriptase
IDF	International Dairy Federation
IFN-γ	interferon-gamma
IL	interleukin
IL-1β	interleukin-1β
iNOS	inducible nitric oxide synthase
ISAPP	International Scientific Association for Probiotics and Prebiotics
KBA	11-keto-β-boswellic acid
LAB	lactic corrosive microbes
LDL-C	low-density lipoprotein-cholesterol
LOX	lipoxygenase
LPO	lipid peroxidation
LTB_4	leukotriene B_4
LTs	leukotrienes
MA	meta-analysis
MB	medulloblastoma

MDA	malondialdehyde
MDR	multidrug-resistant
MFM	Mexican folk medicine
MIC	minimum inhibitory concentration
MPs	medicinal plants
NBCs	natural bioactive compounds
NCEs	new chemical entities
NF-κB	nuclear factor-κB
NIDDM	non-insulin dependent DM
NO	nitric oxide
NP	natural product
NPnEO	non-phenol polyethoxylates
NPs	nanoparticles
$O2^{-\cdot}$	superoxide anion
OH	hydroxyl radical
PAHs	polyaromatic hydrocarbons
PBMCs	peripheral blood mononuclear cells
PBS	phosphate buffer saline
PC	phosphatidylcholine
PCA	principal component
PCBs	polychlorinated biphenyls
PCNA	proliferating cell nuclear antigen
PD	Parkinson's disease
PDI	poly-dispersity index
PGs	prostaglandins
PMNL	polymorphonuclear leukocytes
PTs	pentacyclic triterpenoid
PUFA	poly-unsaturated fatty acid
QPS	qualified presumption of safety
ROO˙	peroxyl radical
ROS	reactive oxygen species
SEM	scanning microscopy
SF5-	pentafluorosulfanyl
SIR-2	silent information regulator-2
SMAC	second mitochondrial-derived activator of caspases
SMAD4	mothers against decapentaplegic homolog 4
SmF	submerged fermentation
SOD	superoxide dismutase

SPR	surface plasmon reverberation
SSF	solid-state fermentation
STLs	sesquiterpene lactones
TBHQ	tert-butyl hydroquinone
TC	total cholesterol
TEM	transmission electron microscopy
TG	triglycerides
TGF-β/RII	transforming growth factor, beta receptor II
THQ	thymohydroquinone
TM	traditional medicine
TNF-α	tumor necrosis factor-alpha
TQ	thymoquinone
ULO	ultra-low oxygen
uMtCK	ubiquitous mitochondrial creatine kinase
VEGF	vascular endothelial growth factor
XRD	x-ray diffraction
ZOI	zones of inhibition

Preface

Green products have been a fertile area of chemical investigation for many years, driving the development of both analytical chemistries and of new synthetic reactions and methodologies.

Green products have always been of key importance to drug discovery, but as modern techniques and technologies have allowed researchers to identify, isolate, extract, and synthesize their active compounds in new ways, they are once again coming to the forefront of drug discovery.

Understanding how to apply the principles of novel green chemistry methods into a vital area such as pharmaceuticals is the overall goal of this volume. It also covers important aspects of green chemistry in the pharmaceutical industry.

Green products chemistry remains a high priority in modern organic synthesis and pharmaceutical research, with important environmental and economic implications. This book presents comprehensive coverage of natural products (NPs) chemistry techniques for organic and medicinal chemistry applications, summarizing the available new technologies, analyzing each technique's features and green chemistry characteristics, and providing examples to demonstrate applications for green organic synthesis and medicinal chemistry.

Applied pharmaceutical science and microbiology is a student-friendly research book to the key areas of chemistry required by all pharmaceutical science students. The book provides a comprehensive overview of the various areas of general, organic, and natural products chemistry.

This volume expands upon presented concepts with the latest research and applications, providing both the breadth and depth researchers need. The book also introduces the topic of natural pharmaceuticals and green microbial technology with an overview of key concepts. Scientists and graduate students will gain a unique insight into green microbial technology, and natural pharmaceuticals today with practical implementation. It also helps students, teachers, researchers, and industrial pharmaceutical scientists use elements of biology, and chemistry in their work and study.

In Chapter 1, *Verbena carolina* is used as a decoction in Mexican folk medicine (MFM) with applications *vs.* digestive problems and for dermatological infections. This work reports high-pressure liquid chro-matography analysis, and extracts and isolated-compounds free radical scavenging capacity, which antimicrobial analyses *vs.* the bacteria *Staphylococcus aureus*, *Enterococcus faecalis*, *Escherichia coli* and *Salmonella typhi*, and the fungi *Candida albicans*, *Trichophyton mentagrophytes* and *T. rubrum* are informed, and aqueous-extracts acute oral toxicity in mice. Major secondary metabolites in *V. carolina* extracts are isolated by conventional phytochemical methods, which consist of three terpenoids and four phenolics. Their contents are determined by chromatography in different samples from dissimilar locations. The results indicate that ursolic acid, hispidulin, verbenaline, hastatoside, verbascoside, hispidulin 7-*O*-β-D-glucuronopyranoside,   and   pectolinaringenin-7-*O*-α-D-glucuronopyranoside are the main constituents, with verbascoside being the most abundant and a significant antioxidant activity in reactive oxygen species (ROS). Hispidulin is the only active compound *vs. T. mentagrophytes* and *T. rubrum*. The aqueous extract shows no significant toxicity. *Mentha* comprises several aromatic species, which are cultivated world-over because of their distinct aroma and commercial value. In addition to traditional food flavoring uses, *Mentha* spp. are well recognized for their folk medicinal services, especially to treat cold, fever, digestive, and cardiovascular disorders.

In Chapter 2, microbial fermentation has been used generally for the conservation of nutrients, the medical assistance of which has become exposed. Fermented diets and beverages are a heterogeneous class of foods with a global significance for the human economy, nutrition, and well being for centuries. In this chapter, we bring up the potential risks for human wellbeing related with uncontrolled nutrition fermentation and we talk about biotechnological approaches ineffective to resolve fermented nutrient protection. The present trend in a situation to manufacture and usage of fermented foods involves consideration of newer tools and tech-niques as well as microorganisms to produce newer products of human interest and to increase the shelf life of the products. One of the most vital foods produced from the microbial fermentation is industrial enzyme. Besides enzymes, sugars, alcohols, vitamins, amino acids, antibiotics, fermentation product itself can be used as food supplements with benefi-cial effects on the consumers, known as probiotics.

Recent reports on imperative medicinal potentials of *Boswellia serrata* are investigated in Chapter 3. *Boswellia serrata* is thought of as an

Ayurvedic medication and intimate Burseraceae class. The plant is cosmopolitan in India and grows in the dry steep forests of provinces such as Rajasthan, Madhya Pradesh, Gujarat, Bihar, Assam, Orissa, etc. Hindus, Babylonians, Persians, Romans, Chinese, Greeks, and Yankee civilizations used it primarily for ritual, embalming, and for its incense in cultural functions. This plant is mentioned in ancient Unani texts as an efficient remedy for respiratory disease, asthma, cough, vessel diseases, diarrhea, dysentery, ringworm, boils, fevers (antipyretic), skin, and blood diseases, mouth sores, channel discharges, etc. The qualitative phytochemical examination of plant extract indicates the presence of tannic acid, pentosans, lignin, holocellulose, β-sitosterol, volatile (cadinene, eleneol, gereniol, linalool, β-pinene, phenols, terpenyl acetate, bornyl acetate, etc.) and non-volatile (diterpene alcohol serratol, α-amyrin, and β-amyrin, triterpenic acids *viz.*, boswellic acid) oils. It primarily possesses anti-arthritic, medicine, anti-hyperlipidemic, anti-cancer, hypoglycemic, anti-asthmatic, analgesic, hepatoprotective, etc. pharmacological activities.

Microbial transformations of natural compounds picked up their significance with the advancement of steroid drugs where such procedures play a vital work. Biotransformation is the premise of life. Microorganisms have been generally related to steroid biotransformation to get ready specific derivatives, the creation of which is troublesome by conventional manufactured strategies. The acknowledgment of microbial biotransformation as a significant assembling device has expanded in pharmaceuticals. The aim of Chapter 4 is to present the generation of natural pharmaceuticals based on the microbial transformation of herbal constituents. Depicted are some effective uses of microbial transformation methods for the generation of steroid drugs or potentially their significant phytoconstituents as well as microbial modifications of terpenes, alkaloids, flavonoids bearing subsidiaries with improved biological activities.

Actinidia deliciosa is also known to be green kiwi, kiwi fruit, Chinese gooseberry, yang tao, etc. which belongs to the sub-family of genus *Actinidia.* Chapter 5 highlighted anti-hypertensive, anti-diabetic, anti-carcinogenic, anti-fungal, hepatoprotective, anti-asthma, anti-platelet, anti-nociceptive, anti-retroviral, etc. of Kiwi fruit. The phytochemicals; amino acids (histidine, arginine, tyrosine, valine, and phenylalanine), vitamins (vitamin B_1 (thiamine), vitamin B_2 (riboflavin), vitamin B_3 (niacin), vitamin B_6 (pyridoxine), vitamin B_9 (folate), vitamin C, vitamin E, and vitamin K), anthocyanins (carotenoids, beta-carotene, and lutein), organic acids (citric acid, quinic acid, and maleic acid), tannins, etc. The other parts of kiwi plant also contain various active constituents like phenolic acids (vanillic acid, hydroxyl cinnamic acid, and caffeic acid), coumarins (umbelliferon and fraxetin), steroids (sitosterol), sesquiterpenoids (alpha-farnesene and germacrene D), carbohydrates (starch, cellulose, pectin, and sugars), minerals (Mg, P, Mn, K, Na, and Zn), protein (actinidin), flavonoids (quercetin and kaempferol), organic acids (citric acid and quinic acid), etc. which are responsible for the pharmacological responses are comprehensively described. The methods and modes of production, storage, side-effects, and traditional uses of kiwi fruits are provided.

Plant-based medicines or herbal drugs have an essential role in developing novel therapeutics for innumerable diseases. Traditional medicines (TM) have a long history date back to thousands of years. Asian countries like India and China have a rich and diverse legacy of traditional systems of medicines like Ayurveda, Siddha, and traditional Chinese medicine. *Ocimum sanctum Linn* also is known as "Holy basil" or "*Tulsi*" is one of the indispensable medicinal plants (MPs) reported in traditional medicines. It has been reported to possess diverse pharmacological activities like anti-microbial, immunomodulatory, anti-stress, anti-inflammatory, anti-ulcer, anti-diabetic, hepatoprotective, chemoprotective, anti-hyperlipidemic, cardioprotective, anti-oxidant, antitussive, radioprotective, memory enhancing, anti-arthritic, anti-fertility, anti-hypertensive, anti-coagulant, anti-cataract, anthelmintic, and anti-nociceptive activities. It has a safe record of human consumption for thousands of years. Hence, the plant has been extensively studied by several investigators and its active constituents are isolated and structure elucidations are done with modern analytical techniques. Some of the individual components were comprehensively studied at the molecular level and the ongoing quest of researchers enabled us to know much about the chemical constitution of *O. sanctum Linn*. Therefore, in Chapter 6 the summarization is done along with structures of phytochemicals of sacred basil with reported biological activities.

Cancer is a category of diseases within which a cell or a bunch of cells displays uncontrolled growth, invasion, and metastasis. Therapy is the use of anti-tumor medicine to treat cancer by busy bodied the expansion ability of cancer cells (CRC). Metal nanoparticles (NPs) have tremendous applications within the space of chemical change, opto-physical science, diagnostic biological probes, etc. Silver (Ag) could be a metallic element and it is applicable in medicines is priceless for millenniums. The preparation of NPs involves main three strategies: (a) physical, (b) chemical, and (c) biological. As there are varying strategies like sol-gel method, chemical precipitation, reverse particle, hydrothermal methodology, and biological strategies, etc. are employed to synthesize silver nanoparticles (AgNPs). The biologically synthesize AgNPs in nanobiotechnology space have enhanced its importance to make eco-friendly; value effective, stable NPs, and their advantages in medicines, agriculture, and physics. AgNPs are synthesized from completely diverse safe plants and can be applied in pharmaceutical and biological fields. The biological strategies are eco-friendly, value-effective, and do not involve the utilization of harmful chemicals. Chapter 7 focuses on the synthesis and characterization of AgNPs fabricated from natural plant and organism extracts for the treatment of cancer.

In Chapter 8, we are focusing upon the nanoparticles based drug delivery system as a promising technology in microbial pharmaceutics and extensive applications thereof. Antimicrobial resistance is a budding problem that has obstructed the world of nearby starting of the end for the old generation antimicrobials. Nanoparticles give a positive response to anti-microbial resistance, which could trigger the innovation and make a new generation of antimicrobials for the treatment in the future using nanotechnology-based drug delivery. Nanotechnology is progressively exhausting technology both in nano-medicine and nano-material for the diagnosing and treatment of innumerable diseases and disorders. Nanoparticles presenting worthy potential in the treatment of bacterial infection, where developed several drug deliveries like; microbial triggered drug delivery, micro-encapsulation, pH-sensitive, surface charge changing nanoparticles for treating various bacterial infection. It will be possible by developing novel formulation into nanotechnologies that can increase the therapeutically longest duration to act against microbial infection. However, many unique physicochemical characteristics of nanoparticles can offer new antibacterial modes of action that can be explored. Antimicrobial nanoparticles and nanosized drug delivery transporters have arisen as effective agents against various microbial infections. Nanoparticles having distinctive properties other than their small and tailored particle size like improved reactivity, functionalizable structure, and appreciable surface area. In contrast to the conventional antimicrobial agents, nanoparticles help in reducing the toxicity, controlling resistance, and reducing the cost. Additionally, nanoparticles are also improving the therapeutic and pharmacokinetic properties of the antimicrobial agents. Nanotechnology also helps in the development of cost-effective, accurate, and fast diagnosis, detection, and management by treating microbial agents. They work as nano-anti microbial agents and their extensive ability of treating diseases and disorders. Besides, nanoparticles use multiple biological pathways to apply their antimicrobial mechanisms like the destruction of the cell wall, DNA inhibition, and enzyme synthesis. Furthermore, the preparation of these nanoparticles is more cost-effective than antibiotics synthesis and also more stable for long term stability and they also withstand to grating condition. Spherulites are small, rounded multilamellar lipidic vesicles that can encapsulate biomolecules and may be used as a carrier for drug delivery. Spherulites resemble liposomes but they are obtained by a simpler process (shearing lamellar phases) and have lamellae up to the very center like an onion. Spherulites offer high encapsulation efficiency, cost-effective method of preparation, uniform structure, enhancement of bioavailability, stability, and high reproducibility of the manufacturing process at an industrial scale over other multilamellar drug delivery systems such as liposomes and niosomes and environment-friendly technology. Spherulites have the ability to incorporate both hydrophilic and lipophilic active molecules without the use of organic solvents. Peptides and protein molecules quickly degraded in the human body, especially in GIT. Therefore, spherulites are used for the delivery of these molecules as their structure allows protecting them from enzymatic degradation. Oral delivery of the anticancer drug is a main concerning pharmacy. The main issues faced are the poor and variable bioavailability of these drugs when administered through the oral route. Spherulites have been used to solubilize highly insoluble anticancer drugs, providing an aqueous pharmaceutical vehicle for these molecules.

Chapter 9 highlights the formation of spherulites, confirmed by light scattering studies which revealed the birefringent nature of spherulites and multilayered structure. The prepared spherulites were evaluated for drug encapsulation efficacy, drug content, particle size, in vitro drug release, and stability as per ICH guidelines. Thus, Chapter 9 gives emphasis on green and sustainable technology as spherulites are considered as vesicular drug delivery system right from dermatological formulation to vaccines.

The role of omega-3 fatty acids in different neurodegenerative disorders is discussed in Chapter 10. Omega-3 fatty acids are related to healthy aging all over life. A moment ago, from fish, source omega-3 fatty acids DHA and EPA have been related to fetal development in AD and cognition related problems. Nevertheless, since the human body does not effectively produce few omega-3 fatty acids, while obtaining from marine sources. The brain is extremely augmented with lipids bilayers. Therefore, it is accountable to undertaken that the combination of different fatty acids within brain ha importance for brain functioning, counting neuropsychiatric and cognition development. To know the special effects form fatty acids ingestion, we required to depend on the mutual evaluation of experimental studies, observations, and interventional studies. The brain is enriched with lipid bilayers of omega-3 fatty acid constituents like DHA and EPA may have lots of effects on the brain. The potential function of omega-3 fatty acid, DHA, and EPA in the inhibition of cognition degeneration including AD has involved key importance from the last few decades. Additionally, AD has delivered the utmost positive sign to backing the situation that inflammation gives to a neurodegenerative disorder. The main etiology of the AD is unidentified, but genetic and ecological parameters are supposed, like family history, poor lifestyle and poor diet, the existence of any chronic disease like cardiovascular disease, increased age, and diabetes. It is assumed that primarily, management or prevention of inflammation may delay the indications of AD. Normally, changes the brain physiology with increased aging, comprising reduction of elongated chain omega-3 fatty acids as well as the brain of AD patients have reduced the DHA presence. Provided supplementary DHA can decrease the indications of inflammations. The health-related benefits of omega-3 fatty acids are really appreciable. Presently, there is no evidence of maintenance or confirmation of that modifiable risk parameters like herbal medicines, nutritional supplements, modified diet, etc. are associated with reduced the risk factors of AD or the cognition failure. Even though, multiple studies are showing that lifestyle and diet-related parameters are the main risk factors for AD and cognition behavior. Chapter 10 basically focused on the different functions of omega-3 fatty acids mainly DHA and EPA in different neurodegenerative diseases and disorders.

Chapter 11 highlights pharmacological effects and molecular targets of thymoquinone (TQ) and enumerates its benefits against various diseases and complications such as oxidative stress, inflammation, cancer, diabetes, cardiovascular problems, and microbial infections. TQ is a dietary component and found in the highest amount in *N. sativa* seeds which are popular among several traditional systems of medicine. Multiple therapeutic potential and low toxicity make *N. sativa* and TQ popular worldwide to be used as a dietary supplement and attracted scientific attention for pharmacological and drug development approaches. Pharmacological properties, high therapeutic index, pharmacokinetics, lipophilicity, efficacy, and low toxicity profile represent TQ as a capable molecule for drug development. Multiple studies described that TQ modulates various receptors, transcription factors, enzymatic, and cell signaling pathways to exhibit its pharmacological effects. It also displayed synergistic effects with many conventional drugs and provides a justification for its use in multifactorial pathogenesis. Further research studies should continue in order to understand the precise molecular mechanism of TQ and to develop its potent analogs. Regardless of its therapeutic efficacy in animal models, long-term clinical trials should be conducted to bring TQ into the clinic.

CHAPTER 1

Antimicrobial, Antioxidant, and Composition of *Verbena carolina* and *Mentha*

FRANCISCO TORRENS[1] and GLORIA CASTELLANO[2]

[1]*Department of Molecular Chemistry, University of Valencia, Valencia, Spain, E-mail: torrens@uv.es*

[2]*Department Experimental Sciences and Mathematics, Catholoc University of Valencia "San Vincent Martir," Valencia, Spain*

ABSTRACT

Verbena carolina is used as a decoction in Mexican folk medicine (MFM) with applications *vs.* digestive problems and for dermatological infections. This work reports high-pressure liquid chromatography analysis, and extracts and isolated-compounds free radical scavenging capacity, which antimicrobial analyses *vs.* the bacteria *Staphylococcus aureus*, *Enterococcus faecalis*, *Escherichia coli,* and *Salmonella typhi*, and the fungi *Candida albicans*, *Trichophyton mentagrophytes*, and *T. rubrum* are informed, and aqueous-extracts acute oral toxicity in mice. Major secondary metabolites in *V. carolina* extracts are isolated by conventional phytochemical methods, which consist of three terpenoids and four phenolics. Their contents are determined by chromatography in different samples from dissimilar locations. The results indicate that ursolic acid, hispidulin, verbena line, hastatoside, verbascoside, hispidulin 7-*O*-β-D-glucuronopyranoside, and pectolinaringenin-7-*O*-α-D-glucuronopyranoside are the main constituents, with verbascoside being the most abundant and significant antioxidant activity in reactive oxygen species (ROS). Hispidulin is the only active compound *vs. T. mentagrophytes* and *T. rubrum*. The aqueous extract shows no significant toxicity. *Mentha* comprises several aromatic species, which are cultivated world-over because of their distinct aroma and commercial value. In addition to traditional food flavoring uses, *Mentha* spp. are well recognized for their folk medicinal services, especially to treat cold, fever, digestive, and cardiovascular disorders.

1.1   INTRODUCTION

Phycotoxins chemistry and biochemistry were reviewed [1]. Cosmetic pharmaceuticals (*cosmeceuticals*) and active cosmetics were revised [2]. Pure total flavonoids from citrus improve non-alcoholic fatty liver disease regulating the toll-like receptor/C-C chemokine ligand signaling pathway [3]. Post-emergent herbicidal activity of *Eucalyptus globulus* essential oil (EO) was reported [4]. Oregano, rosemary, and thyme EOs were published as quality indicators of spices in supermarkets [5]. The EOs was reviewed as antimicrobial agents [6]. Botanical sources, chemistry, analysis, and bioactivity of furanocoumarins of pharmaceutical interest were revised [7]. Prediction of vascular endothelial growth factor-C was informed as a key target of pure total flavonoids from *Citrus vs.* non-alcoholic fatty liver disease in mice *via* net pharmacology [8]. The phytochemistry and pharmacology of *Rubus chingii* Hu was discussed [9].

 Verbena carolina L. (Verbenaceae) is one predominant *Verbena* spp. that grows in America, and widely distributed from Southwestern US to Nicaragua in Central America. In Mexican folk medicine (MFM), it is popularly known as *verbena, ajenjo grande, hierba de San José, nardo de campo, Santa María, poleo negro* and *wahichuri* (Tarahumara). It is generally referred to as *vervain*. It presents a traditional medicine (TM)-efficacy history in MFM. Most of the plant, except for the roots, is used as a decoction in MFM with applications *vs.* diarrhea, vomit, and dysentery, or as a purgative. A decoction of its aerial parts is used to dissolve bladder stones, as a diuretic, and to treat wounds, dandruff, allergies, and dermatitis. It is one of the constituents of a skincare preparation that shows melanogenesis suppression. A verbena-extracts protective effect in CCl_4-induced rat liver injury was informed, and its chemical nutrients, toxic factors, and digestibility were reported. Notwithstanding, the chemical and bioanalyses related to its TM uses and safe prescription are scarce. It was selected for a program that aims to characterize and include MFM flora in Mexican Herbal Pharmacopoeia. Genus *Mentha* comprises several aromatic species, which are cultivated world-over because of their distinct aroma and commercial value. In addition to traditional food flavoring uses, *Mentha* spp. is well recognized for their folk medicinal services, especially to treat cold, fever, digestive, and cardiovascular diseases (CVDs). Bioinhibitory activities were ascribed to *Mentha*. *Mentha*-herbs TM uses and bioactivities can be linked to the occurrence of a wide array of natural bioactive compounds (NBCs). As a source of NBCs, *Mentha* herbs can be explored as a promising candidate for the development of nutrapharmaceuticals.

Earlier publications in *Nereis, etc.* classified yams [10], lactic acid bacteria [11], fruits [12], food spices [13], and oil legumes [14] by principal component (PCA), cluster (CA), and meta-analysis (MA). The molecular classifications of 33 phenolic compounds derived from the cinnamic and benzoic acids from *Posidonia oceanica* [15], 74 flavonoids [16], 66 stilbenoids [17], 71 triterpenoids and steroids from *Ganoderma* [18], 17 isoflavonoids from *Dalbergia parviflora* [19], 31 sesquiterpene lactones (STLs) [20, 21] and STL artemisinin derivatives [22] were informed. A tool for interrogation of the macromolecular structure was reported [23]. Mucoadhesive polymer hyaluronan favors transdermal penetration absorption of caffeine [24, 25]. Polyphenolic phytochemicals in cancer prevention and therapy, bioavailability, and bioefficacy were reviewed [26]. From Asia to the Mediterranean, soya bean, Spanish legumes, and commercial *soya bean* PCA, CA, and MA were published [27]. The natural product (NP) antioxidants (AOs) from herbs and spices improved the oxidative stability and frying performance of vegetable oils [28]. The relationship between vegetable oil composition and oxidative stability was revealed *via* a multifactorial approach [29]. It was informed chemical and biological screening approaches to phytopharmaceuticals [30], cultural interbreeding in indigenous and scientific ethnopharmacology [31], ethnobotanical studies of medicinal plants (MPs), underutilized wild edible plants, food, medicine [32], biodiversity as a source of drugs, *Cordia, Echinacea, Tabernaemontana, Aloe* [33], immunomodulatory molecules from Himalayan MPs [34], phylogenesis by information entropy, avian birds and 1918 influenza virus [35]. The aim of this work is to review the antimicrobial, AO activities (AOAs), and high-performance liquid chromatography (HPLC) determination of the major components of *Verbena carolina*, and *Mentha*, a genus rich in vital nutra-pharmaceuticals.

1.2   POPULAR, TRADITIONAL, AND MODERN MEDICAL MATTER

In the history of drugs therapeutics, three fundamental stages are usually distinguished [36]. The first corresponds to primitive communities and archaic cultures. It was characterized by the merely empirical use of drugs substances, combined with magical and religious credence interpretations. It remained to this day in the folk medicine of the developed societies. The second stage covers from the beginning of scientific medicine in classical Greece (5th century) to the first decades of the 19th century. Its basics were of rational type. It was limited to natural healing products, mainly vegetables, and for that reason, it relied on natural history, especially in botanic. In the classic Antiquity, the products were practically limited to those of the Mediterranean area. During the Middle Ages, their number rose noticeably, especially because of the contributions of Islamic physicians. Renaissance geographical discoveries enriched the arsenal with products from America and the Far East. The development of modern chemistry made possible the third stage. It started with the analysis of natural medicine and the isolation of its active chemical principles, continued with the constitution of experimental pharmacology as an institutionalized subject, and culminated with chemotherapy.

1.3   ANTIMICROBIAL, ANTIOXIDANT ACTIVITIES, AND COMPOSITION OF *VERBENA CAROLINA*

Verbena carolina-major-components antimicrobial, AOAs, and HPLC determination were informed [37]. Ursolic acid (**1**), hispidulin (**2**), verbenalin (**3**), hastatoside (**4**), verbascoside (**5**), hispidulin 7-*O*-β-D-glucuronopyranoside (7OβGH, **6**) and pectolinaringenin-7-*O*-α-D-glucuronopyranoside (**7**) isolation and identification from its extracts were achieved. Major secondary metabolites in *V. carolina* extracts are isolated by conventional phytochemical methods, which consist of three terpenoids (**1**, **3**, **4**) and four phenolics (**2**, **5–7**). Except for hispidulin (**3**), neither the aqueous extract (CAqE), nor compounds (**5**), (**6**) or (**7**) were active *vs.* the tested microorganisms. Isolated-compounds structures are shown (*cf.* Figure 1.1).

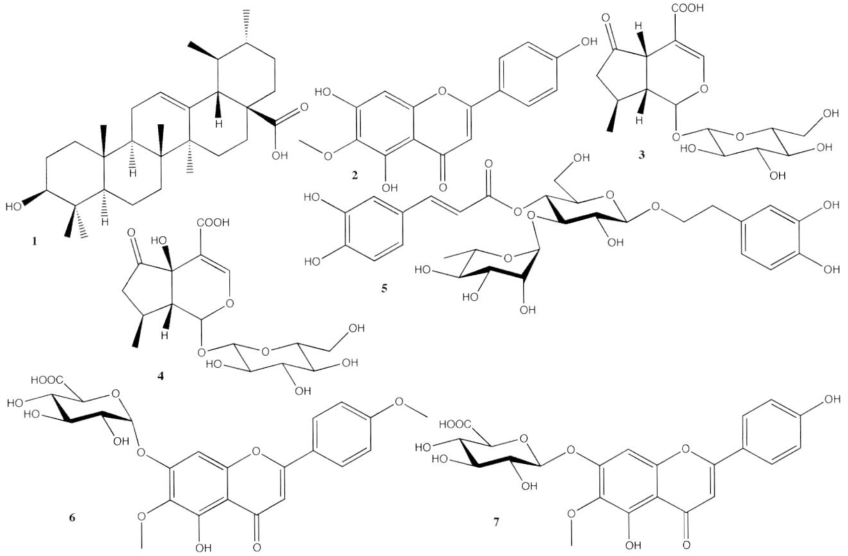

FIGURE 1.1   Compounds from *V. carolina*: ursolic acid (**1**); hispidulin (**2**); verbenaline (**3**); hastatoside (**4**); verbascoside (**5**); hispidulin 7OβGH (**6**); pectolinaringenin-7-*O*-α-D-glucuronopyranoside (**7**).

1.4   *MENTHA*: A GENUS RICH IN VITAL NUTRAPHARMACEUTICALS

Mentha is a genus rich in vital nutrapharmaceuticals [38]. Figure 1.2 shows the structure of some common chemical constituents of *Mentha* EOs.

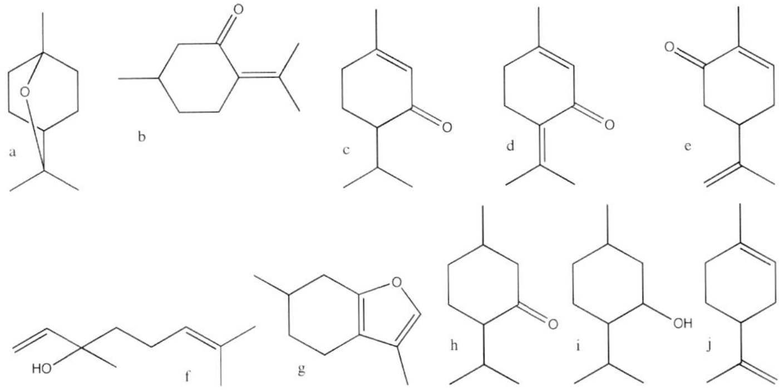

FIGURE 1.2   (a) Cineole; (b) pulegone; (c) piperitone; (d) piperitenone; (e) carvone; (f) linalool; (g) menthofuran; (h) menthone; (i) menthol; (j) limonene.

1.5   DISCUSSION

Since the history of mankind, plants not only provided shelter, fuel, and food but also they were always a vital source of TM imparting health benefits. Because of the high cost and safety of pharmaceutical medicine, renewed interest exists in the use of plants/herbs as food and TM. The medicinal NPs obtained from a number of plants are considered safer. 80% of the world's population, especially in Africa and South Asia, depends on TM, mainly the plant-derived NPs and phytomedicines, to accomplishing their basic healthcare needs. Many food and MPs, especially aromatic herbs and spices, were evaluated for nutrapharmaceutical potential, because of containing a wide array of NBCs and new chemical entities (NCEs).

Traditionally, the plants of genus *Mentha*, and their extracts and EOs were used for food flavoring and folk TM purposes. Because of the wide range of medicinal properties, the plants resulted in positioned for the treatment of various diseases (e.g., cold, cough, asthma, gastrointestinal problems). The herbs, because of containing a variety of secondary metabolites (e.g., volatile oil, tannins, phenols, flavonoids), exhibit multiple bioactivities (e.g., antibacterial, antifungal, antiviral, urease inhibitory, antihypertensive, antidiarrheal, anti-ulcer, anti-inflammatory, biopesticidal). *Mentha* EOs mainly contained volatile NBCs (e.g., menthol, menthone, menthofuran, piperitenone, piperitone, carvone, pulegone, -pinene, linalool). Anti-inflammatory drugs cause gastric ulcers as a serious adverse effect. *Mentha* spp. possesses anti-inflammatory activity but is unlikely to produce ulcer, as it contains anti-ulcerogenic constituent(s); it offers a huge potential for NPs to be developed as anti-inflammatory agents. Although few reports on the anti-*Helicobacter pylori* activity of *Mentha* spp. are available, need exists to screen *Mentha* spp. for the exploration of NP anti-ulcer agents to treat gastric ulcer disorders. *Mentha longifolia* possesses Ca^{2+} antagonist activity, which can explain many medicinal uses of the plant (e.g., in gastric ulcer, asthma, diarrhea, CVDs); the scope of testing *Mentha* spp. exists for the possible presence of Ca^{2+} antagonist activity and identifying the chemical constituent(s) responsible for the useful activity.

Mentha plant extracts and EOs are in use as ingredients of several foods, herbal cosmetics, skin and hair care products, etc.; however, need exists for standardization of such formulations and NPs for safer human uses. *Mentha* extracts, demonstrating the established antimicrobial activity, can be explored as a natural platform of lead molecules to discovering novel antimicrobial agents to cope with microbial resistance to antibiotics. A need exists to investigate the bioactivities of unexplored *Mentha* spp. and the need exists to explore potential applications in the pharmaceutical industry, and study the mechanisms of action of *Mentha* plant extracts and EOs *via in vivo* models and clinical trials for optimization and standardization of dosages for a specific application.

1.6 FINAL REMARKS

From the present results and discussion, the following final remarks can be drawn.

1. *Verbena carolina*-aqueous-extract beneficial effect, for dermatological conditions, is justified by the results obtained in the antifungal and antioxidant assays. With respect to antimicrobial activity, hispidulin was active *vs.* the fungi *Trichophyton mentagrophites*, *T. rubrum*, and *Candida albicans*. Regarding the scavenging effect, generally individual compounds were more effective than extracts. The results support the plant widespread use in TM.
2. A simple and reliable reversed phase-high-pressure liquid chromatography analytical method was developed and validated for verbena line, hastatoside, verbascoside, and hispidulin 7-*O*-β-D-glucuronopyranoside quantification in *V. carolina*. The method was used to establish markers seasonal and geographical variations in *V. carolina* from several Mexico regions. Its marker compounds must be useful for quality control procedures, which are needed for MPs authentication protocols.

ACKNOWLEDGMENTS

The authors thank support from Generalitat Valenciana (Project No. PROMETEO/2016/094) and Universidad Católica de Valencia San Vicente Mártir (Project No. 2019-217-001).

KEYWORDS

- **antioxidant capacity**
- **essential oil**
- **high-performance liquid chromatography**
- **medicinal plant**
- **meta-analysis**
- **natural bioactive compounds**
- **sesquiterpene lactones**

REFERENCES

1. Botana, L. M., (2007). *Phycotoxins: Chemistry and Biochemistry*. Blackwell: Ames, IA.
2. Sivamani, R. S., Jagdeo, J., Elsner, P., & Maibach, H. I., (2015). *Cosmeceuticals and Active Cosmetics*. CRC: Boca Raton, FL.
3. Wu, L., Yan, M., Jiang, J., He, B., Hong, W., & Chen, Z., (2017). Pure total flavonoids from citrus improve non-alcoholic fatty liver disease by regulating TLR/CCL signaling pathway: A preliminary high-throughput *omics* study. *Biomed. Pharmacother., 93*, 316–326.
4. Ibáñez, J. M. D., & Blázquez, F. M. A., (2018). Post-emergent herbicidal activity of *Eucalyptus globulus* labill. essential oil. *Nereis, 10*, 25–36.
5. Ibáñez, M. D., & Blázquez, M. A., (2019). Essential oils: Quality indicators of spices in supermarkets. *Nereis, 11*, 39–50.
6. Winska, K., Maczka, W., Lyczko, J., Grabarczyk, M., Czubaszek, A., & Szumny, A., (2019). Essential oils as antimicrobial agents: Myth or real alternative? *Molecules, 24*, 2130-1-21.
7. Bruni, R., Barreca, D., Protti, M., Brighenti, V., Righetti, L., Anceschi, L., Mercolini, L., Benvenuti, S., Gattuso, G., & Pellati, F., (2019). Botanical sources, chemistry, analysis, and biological activity of furanocoumarins of pharmaceutical interest. *Molecules, 24*, 2163-1-25.
8. Hong, W., Li, S., Wu, L., He, B., Jiang, J., & Chen, Z., (2019). Prediction of VEGF-C as a key target of pure total flavonoids from *Citrus* against NAFLD in mice via network pharmacology. *Front. Pharmacol., 10*, 582–1–14.
9. Yu, G., Luo, Z., Wang, W., Li, Y., Zhou, Y., & Shi, Y., (2019). *Rubus chingii* Hu: A review of the phytochemistry and pharmacology. *Front. Pharmacol., 10*, 799–1–22.
10. Torrens-Zaragozá, F., (2013). Molecular categorization of yams by principal component and cluster analyses. *Nereis, 2013*(5), 41–51.
11. Torrens-Zaragozá, F., (2014). Classification of lactic acid bacteria against cytokine immune modulation. *Nereis, 2014*(6), 27–37.

12. Torrens-Zaragozá, F., (2015). Classification of fruits proximate and mineral content: Principal component, cluster, meta-analyses. *Nereis, 2015*(7), 39–50.
13. Torrens-Zaragozá, F., (2016). Classification of food spices by proximate content: Principal component, cluster, meta-analyses, *Nereis, 2016*(8), 23–33.
14. Torrens, F., & Castellano, G., (2014). From Asia to Mediterranean: Soya bean, Spanish legumes, and commercial *soya bean* principal component, cluster and meta-analyses. *J. Nutr. Food Sci., 4*(5), 98–98.
15. Castellano, G., Tena, J., & Torrens, F., (2012). Classification of polyphenolic compounds by chemical structural indicators and its relation to antioxidant properties of *Posidonia oceanica* (L.) Delile. *Match. Commun. Math. Comput. Chem., 67*, 231–250.
16. Castellano, G., González-Santander, J. L., Lara, A., & Torrens, F., (2013). Classification of flavonoid compounds by using entropy of information theory. *Phytochemistry, 93*, 182–191.
17. Castellano, G., Lara, A., & Torrens, F., (2014). Classification of stilbenoid compounds by entropy of artificial intelligence. *Phytochemistry, 97*, 62–69.
18. Castellano, G., & Torrens, F., (2015). Information entropy-based classification of triterpenoids and steroids from *Ganoderma. Phytochemistry, 116*, 305–313.
19. Castellano, G., & Torrens, F., (2015). Quantitative structure-antioxidant activity models of isoflavonoids: A theoretical study. *Int. J. Mol. Sci., 16*, 12891–12906.
20. Castellano, G., Redondo, L., & Torrens, F., (2017). QSAR of natural sesquiterpene lactones as inhibitors of Myb-dependent gene expression. *Curr. Top. Med. Chem., 17*, 3256–3268.
21. Torrens, F., & Castellano, G. (2018). Structure-activity relationships of cytotoxic lactones as inhibitors and mechanisms of action. *Curr. Drug Discov. Technol.* doi: 10.2174/1570163816666190101113434.
22. Torrens, F., Redondo, L., & Castellano, G., (2017). Artemisinin: Tentative mechanism of action and resistance. *Pharmaceuticals, 10*, 20-4-4.
23. Torrens, F., & Castellano, G., (2014). A tool for interrogation of macromolecular structure. *J. Mater. Sci. Eng. B, 4*(2), 55–63.
24. Torrens, F., & Castellano, G., (2014). Mucoadhesive polymer hyaluronan as biodegradable cationic/zwitterionic-drug delivery vehicle. *ADMET DMPK, 2*, 235–247.
25. Torrens, F., & Castellano, G., (2015). Computational study of nano-sized drug delivery from cyclodextrins, crown ethers, and hyaluronan in pharmaceutical formulations. *Curr. Top. Med. Chem., 15*, 1901–1913.
26. Estrela, J. M., Mena, S., Obrador, E., Benlloch, M., Castellano, G., Salvador, R., & Dellinger, R. W., (2017). Polyphenolic phytochemicals in cancer prevention and therapy: Bioavailability *versus* bioefficacy. *J. Med. Chem., 60*, 9413–9436.
27. Torrens, F., & Castellano, G., (2014). From Asia to Mediterranean: Soya bean, Spanish legumes and commercial *soya bean* principal component, cluster and meta-analyses. *J. Nutr. Food Sci., 4*(5), 98–98.
28. Redondo-Cuevas, L., Castellano, G., & Raikos, V., (2017). Natural antioxidants from herbs and spices improve the oxidative stability and frying performance of vegetable oils. *Int. J. Food Sci. Technol., 52*, 2422–2428.

29. Redondo-Cuevas, L., Castellano, G., Torrens, F., & Raikos, V., (2018). Revealing the relationship between vegetable oil composition and oxidative stability: A multifactorial approach. *J. Food Compos. Anal., 66*, 221–229.

30. Torrens, F., & Castellano, G. (2019). Chemical/biological screening approaches to phytopharmaceuticals. In: Pourhashemi, A., Deka, S. C., & Haghi, A. K., (eds.), *Research Methods and Applications in Chemical and Biological Engineering*. Apple Academic-CRC: Waretown, NJ.

31. Torrens, F., & Castellano, G. (2019). Cultural interbreeding in indigenous/scientific ethnopharmacology. In: Pourhashemi, A., Deka, S. C., & Haghi, A. K., (eds.), *Research Methods and Applications in Chemical and Biological Engineering*. Apple Academic-CRC: Waretown, NJ.

32. Torrens, F., & Castellano, G. (2019). Ethnobotanical studies of medicinal plants: Underutilized wild edible plants, food, and medicine. In: Haghi, A. K., (ed.), *Innovations in Physical Chemistry*. Apple Academic-CRC: Waretown, NJ.

33. Torrens, F., & Castellano, G. (2019). Biodiversity as a source of drugs: Cordia, Echinacea, Tabernaemontana and aloe. In: Aguilar, C. N., Ameta, S. C., & Haghi, A. K., (eds.), *Green Chemistry and Biodiversity: Principles, Techniques, and Correlations*. Apple Academic-CRC: Waretown, NJ.

34. Torrens, F., & Castellano, G. (2020). Immunomodulatory molecules from Himalayan medicinal plants. In: Esteso, M. A., Ribeiro, A. C. F., & Haghi, A. K., (eds.), *Chemistry and Chemical Engineering for Sustainable Development: Best Practices and Research Directions*. Apple Academic-CRC: Waretown, NJ.

35. Torrens, F., & Castellano, G., (2008). Phylogenesis by information entropy: Avian birds and 1918 influenza virus. In: Turner, S., & Yunus, J., (eds.), *Modeling and Simulation* (Vol. 2, pp. 1, 2). IEEE: London.

36. López, P. J. M., Bujosa, H. F., & Micó, N. J., (1980). *Guía de la Exposición Historicomédica 15 Abril/15 Mayo 1980*. Universitat de València: València, Spain.

37. Lara-Issasi, G., Salgado, C., Pedraza-Chaverri, J., Medina-Campos, O. N., Morales, A., Águila, M. A., Avilés, M., et al., (2019). Antimicrobial, antioxidant activities, and HPLC determination of the major components of *Verbena carolina* (Verbenaceae). *Molecules, 24*(10), 1970; https://doi.org/10.3390/molecules24101970.

38. Anwar, F., Abbas, A., Mehmood, T., Gilani, A. H., & Rehman, N. U., (2019). *Mentha*: A genus rich in vital nutra-pharmaceuticals: A review. *Phytother. Res., 33*(10), 2548–2570. doi: 10.1002/ptr.6423 (Epub 9 July 2019).

CHAPTER 2

Recent Trends in Microbial Fermentation

AKSHADA ATUL BAKLIWAL, SWATI GOKUL TALELE, and
ANIL G. JADHAV

*Sandip Institute of Pharmaceutical Sciences, Nashik, Maharashtra,
India, E-mail: swatitalele77@gmail.com (S. G. Talele)*

ABSTRACT

Microbial fermentation has been used generally for the conservation of nutrients, the medical assistance of which has since become exposed. Fermented diets and beverages are a heterogeneous class of foods with a global significance for the human economy, nutrition, and well being for centuries. In this chapter, we bring up the potential risks for human well-being related with uncontrolled nutrition fermentation and we talk about biotechnological approaches ineffective to resolve fermented nutrient protection. The present trend in a situation to manufacture and usage of fermented foods involves consideration of newer tools and techniques as well as microorganisms to produce newer products of human interest and to increase the shelf life of the products. One of the most vital foods produced from the microbial fermentation is industrial enzyme. Besides enzymes, sugars, alcohols, vitamins, amino acids, antibiotics, fermentation product itself can be used as food supplements with beneficial effects on the consumers, known as probiotics.

2.1 INTRODUCTION

Fermentation has been broadly utilized for the creation of a wide assortment of substances that are exceptionally helpful to people

[8]. Fermentation, with its wide exhibit of use and huge advantages, has created an enormous spurt in the use of microorganisms for the mechanical generation of fermented products. Instead of the fact that fermented products are being utilized from ancient periods in allover this planet, because of the varieties of various fermentation strategies and generation of a great deal of auxiliary bioactive mixes, broad research is going on varieties of fermentation methods and investigation of fresher strains of fermenting microorganisms and the molecular occasion in that.

Various fermented items like proteins, sugars, alcohols, and other worth included materials can be acquired as fermented products and microbial fermentation can be embraced as a condition neighborly clean technology for the creation of commodities of significance to chemical, energy, and food industries [9]. In the present time, an expanding overall enthusiasm for elective wellsprings of vitality not just because of the consistent exhaustion of restricted petroleum product stock yet additionally for protected and better condition and biomass-based fuel advancement innovations using microbial extracellular enzymatic hydrolysis and fermentation is picking up force. Since the major compel in the commercialization of fermented products lies in the staggering expense of the costly substrates utilized, interests have been produced in the usage of rural squanders rather than unadulterated substrates in the fermentation medium. Besides, various biotechnological methodologies have been embraced in particular varieties in development medium, immobilization, strain improvement through site guided or old-style mutagenesis to upgrade the creation of the bioactive mixes of intrigue. Advances in microbiology and fermentation innovation have proceeded relentlessly up until the present.

The fermentation of nourishment by microorganisms has been utilized for centuries as a procedure to guarantee an expanded time span of usability and improve the usefulness, surface, and kind of food items. The primary proof of dairy fermentation exists from around 7,000 years back, wherein earlier Europeans produced cheese. Techniques have advanced from unconstrained fermentation by the indigenous microbial populace to pre-determination of starter societies with known qualities. Lactic corrosive microbes (LAB) are the significant microscopic organisms utilized in food fermentation around the world. LAB comprise of a horde of genera including, yet not limited to, *Lactobacillus, Lactococcus, Streptococcus,*

Leuconostoc, *Pediococcus*, and *Enterococcus* [10–12]. Despite the fact that the LAB is a different group of bacteria, many species enjoy historical "generally regarded as sheltered" (GRAS) and "qualified presumption of safety" (QPS) status by the Food and Drug Administration (FDA) and European Food Safety Authority (EFSA), respectively. LAB aging has for some time been perceived to give useful impacts on human wellbeing through the tweak of the intestinal microbiota. These either directly or indirectly affect the host microbiota, which in turn can prompt an impact on wellbeing. The utilization of these microorganisms in fermentations to deliver functional foods has significantly expanded as of late. The utilization of fermented foods has been related to a scope of medical advantages from infection aversion to upgrading the bioregulation of social issues, for example, stress and anxiety. While the utilization of conventional fermented foods in societies around the globe is accepted to have valuable impacts, not these nourishments have been exposed to suitable preliminaries in which these convictions could be credited or ruined. The advantages of fermented dairy items are being inquired about widely in parts of the world, however, other ethnic fermentations are additionally starting to be contemplated in more detail. These generally fermented foods use uncharacterized starter societies that could have novel properties or be helpful in different fermentations. The potential utilization of microbial fermentation is tremendous both in wellbeing and in biotechnology and will be a significant zone of research and creation in the coming decades [13–15]. Yeasts and molds are likewise noticeable fermented living beings in alcoholic and certain cheese fermentation (Figure 2.1).

The present overview manages the continuous change in the patterns and use of fermentation methods and fermented products all through the world.

2.1.1 MECHANISM OF FERMENTATION

Fermentation is the synthetic change of natural substances into more straightforward mixes by the activity of compounds, complex organic catalysts, which are delivered by microorganisms, for example, molds, yeasts, or microscopic organisms. It might be oxygen-consuming or anaerobic in nature. It is normally viewed as an anaerobic procedure whereby sugar is changed over into corrosive or liquor. Salt assumes a critical job in conventional fermentation by making conditions that support the microscopic organisms, counteracting the development of pathogenic microorganisms, pulling water and supplements from the substrate, and including flavor.

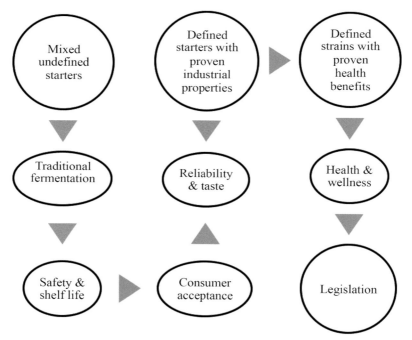

FIGURE 2.1 Flowchart representing the relationships between consumers, fermenting microbes, and dairy products.

Most fermentation is actuated by molds, yeasts, or microscopic organisms, working independently or together. The microbial fermentation might be cultivated either in the submerged state (SmF) or in solid-state (SSF) conditions. Submerged fermentation (SmF) uses free streaming fluid substrates, for example, corn soaks alcohol, molasses, and supplement juices. The created catalysts and bioactive mixes are gotten from fermented stock. This fermentation procedure is most appropriate for microorganisms, for example, microscopic organisms that require high dampness. Then again, strong state fermentation (SSF) holds the gigantic potential for the creation of compounds since

it brings about higher convergence of the items, less effluent generation, and requires moderately basic fermentation types of gear. It is
a three-stage heterogeneous procedure, including strong, fluid, and
vaporous stages, which offers potential advantages for the microbial
development   for bioprocesses and items advancement for which
it has increased noteworthy consideration for the improvement of
modern bioprocesses in the course of the most recent two decades. The
fermentation procedure might be unconstrained or prompted. It is less
vigorously proficient than oxidative phosphorylation since rather than
36, as in oxygen-consuming breath, just 2 ATP s are produced from 1
atom of glucose in fermentation.

2.1.2   ADVANTAGES OF FERMENTED FOODS [16–19]

- **It Re-Establishes Legitimate Microscopic Organisms Balance
  in the Digestive Organs:** Probiotics are mostly parts of some
  portion of a gathering of microscopic organisms that produce lactic
  corrosive and are found in fermented milk, yogurt, and different
  nourishments that have experienced the fermentation procedure.
  The consuming such nourishments can likewise improve the
  bioavailability of supplements, limit the side effects of lactose, and
  decrease the commonness of hypersensitivity in the individuals
  who are defenseless.
- **It Improves the Immune System:** Fermenting the eating foods
  can make the digestive system more grounded and this is less in
  danger for intestinal diseases. Kefir is effectively processed yet it
  additionally colonizes the digestive organs with microorganisms
  that help keep up a healthy immune response. Kefir has been
  utilized to treat tuberculosis and malignancy, as per the United
  Nations University. 80% of your immune system dwells in your
  gut. A solid gut is equal to a healthy immune system. Adding
  fermented foods to your eating routine will help guarantee a
  healthy gut.
- **Liver Protection:** Fermented foods could help individuals with
  Hepatitis C and liver issues. Fermented brown rice may moreover
  help limit the activity of complimentary radicals that can hurt the
  liver and cause or aggravate the advancement of hepatitis and

other liver issues. Consuming fermented brown rice can help limit
the danger of creating severe hepatitis:

o Fermenting products makes them taste extraordinary: like
  miso, kimchi, wine, etc.
o Sourdough bread, modern fermentation technology offers
  broad advantages.
o Fermentation permits vitality creation without oxygen, which
  can be misused to make bread and a few drinks, and enable
  people to keep running for longer periods of time.
o Fermented foods keep significantly longer than new ones.
  That cabbage will turn sour following seven days, yet the
  sauerkraut will keep for months.

2.2 HISTORY OF FERMENTED PRODUCTS

The earliest proof of alcoholic refreshment produced using the organic
product, rice, and nectar, dates from 7000–6600 BCE, in the Neolithic
Chinese town of and winemaking, dates from 6000 BCE, in Georgia,
in the Caucasus region [1]. Old containers containing the remaining
parts of wine have been exhumed in the Zagros Mountains in Iran [2].
There is solid proof that individuals were maturing drinks in Babylon
around 3000 BC [3] pre-Hispanic Mexico around 2000 BC, and Sudan
around 1500 BC. Connections between fermented nourishments
and wellbeing can be followed as far back as old Rome and China,
and remain a region of the incredible enthusiasm for specialists in
present-day times. Wang and Hesseltine opined that most likely the
principal fermentation was found inadvertently when salt was joined
with the nourishment material, and the salt is chosen certain innocuous
microorganisms to make the matured item nutritious and satisfactory
sustenance. Lactic fermentation is an antiquated strategy utilized by
numerous individuals all through the world for saving vegetables [4].

In some instances, fermentation is a vital part of food safety further
than preservation. The procedure was made a stride further by the early
Chinese who originally utilized parasitic compounds to create salt-
matured soya substances, for example, miso, soy sauce, soy pieces, and
fermented tofu. Asian developments specifically have a past filled with
maturing a wide assortment of nourishments fundamentally from East

Asia like Japanese natto (soybeans), Vietnamese mám (fish), Chinese douchi (dark beans), Lao dad daek (fish sauce), Korean banchan for their regular cooking. Fermentation substances were likewise utilized in Eastern societies for therapeutic purposes and for the creation of brew, wine, and raised bread and cheeses. These were before long pursued by East Asian fermented nourishments, yogurt, and other matured milk products, pickles, sauerkraut, vinegar (soured wine), spread, and various customary mixed refreshments. In West African nations, garri was a significant fermented nourishment source from cassava. Different sustenance, for example, the Tanzanian fermented slop togwa, has been found to secure against foodborne diseases in areas that have poor sanitation. In East Asia however customary matured nourishments assumed an uncommonly broad job in their sustenance framework yet these were amazingly best in class and advanced yet the dairy items and other creature proteins (aside from fish) were not widely used [5–7].

2.3   MAKES FOOD MORE DIGESTIBLE [20, 21]

Since the microorganisms predigest the nourishment, the subsequent product is simpler to process. On the off chance that you experience difficulty processing crude leafy foods, fermentation might be useful for you.

- **Crammed with Great Microbes (Probiotics):** A few specialists state that every little 1/2 cup serving of fermented foods can contain up to 10 trillion probiotic life forms. Fermenting separates things which are hard to process and makes a few nourishments increasingly nutritious: like sauerkraut (fermented cabbage) and miso (fermented soybeans).
- **Help Your Intestinal Flora:** Probiotics are live microorganisms that present in your intestinal tracts. Anti-infection agents, notwithstanding an unfortunate eating routine, can execute these organisms, bringing about inconveniences, for example, vaginal diseases and the runs and other gastrointestinal issues. One answer for this is to eat probiotic-rich fermented nourishments, for example, kefir—a sharp yogurt-like beverage-or certain sort of yogurt.

- **It Improves Heart Wellbeing:** There are some milk products that have experienced fermentation useful for the heart. There is proof to demonstrate that fermented milk products can gently diminish truly elevated pulse (likewise known as hypertension).
- **Fermented Nourishments have more Supplements:** The microscopic organisms in fermented nourishments produce more nutrients and supplements as they digest the sugars especially produce B nutrients and vitamin K2.
- **Builds Flavor of Foods:** Fermentation includes another profundity of flavor to fermented foods. They are delicious.
- **Acid Promotes Development of Healthy Microbes in the Gut:** The lactic acid produced during fermentation helps solid microscopic organisms officially present in your gut to multiply. This prompts better gut wellbeing.
- **Enables Check to Sugar Longings:** By adding fermented nourishments to your eating routine, you can constrain, if not totally stop, your sugar desires.

2.4   DISADVANTAGES OF FERMENTED FOODS [22, 23]

- **Risk of Botulism Contamination:** Foods fermented in your home position a threat of botulism tainting. In 2001, a botulism flare-up in an Alaskan town activated 13 people to be hospitalized, with among them enduring heart assault and one requiring a tracheotomy. The scene was brought about by expending fermented beaver tail and paw, a customary neighborhood delicacy. It is connected with the improvement of gastric malignancy: The utilization of fermented foods can likewise help the safe system as observed with kefir, acidic refreshment made by fermenting milk with grains. The impacts of fermented and non-fermented soy foods utilization on the danger of gastric malignancy advancement. The investigation demonstrated that a high admission of fermented soy nourishments expanded the danger of gastric malignant growth while an eating routine that was high in non-fermented soy foods decreased the danger of gastric disease. Locally acquired things lose gainful microscopic organisms: Fermented sustenance sold in numerous stores are handled uniquely in contrast to those that are generally fermented. They have an excess of acid and have been sanitized so they don't ruin immediately. Research has likewise demonstrated that fermented cheese products contain an excessive amount of salt than water. The procedure is, for the most part, biochemical and thus the rate is administered by the life forms included:

1. The waste items may not be anything but difficult to hold up under everlastingly, which is the reason running for a really long time utilizing this procedure causes the development of a synthetic in the muscles that reason weariness, and must be separated while breathing.
2. Alcohol is incredible for restraining bacterial development, yet here and there fermentation goes somewhat wild and you need to water down the mead you made on the grounds that it possesses a flavor like you got it from the alcohol store. Barely healthy for daily use.
3. In people, not all parts of the body ferment, which is the reason the mind goes out from an absence of oxygen before the muscles.
4. Finally, vitality is much simpler to deliver when oxygen is involved.
5. Anaerobic conditions are required alongside extremely exact control of pH and other system conditions. Well-structured fermenters in this manner could be costly on Capex.
6. In cases like yeast, the greatest item fixation that the way of life can shoulder is about 15% and no more.
7. Fermented foods are insufficient; however, they have destructive properties too.

2.5   SIDE EFFECTS OF FERMENTATION

- When food ferment, or deteriorate, certain waste products are created by the microorganisms which separate the nourishment. One of these results is liquor. Many fermented nourishments, for example, soy sauce, contain a lot of liquor. The liquor in fermented foods is typically a little amount, however even limited quantities of liquor influence the cells of the body.

- Another acid that outcomes from fermenting are lactic acid. Lactic acid is a waste item. In the event that you have ever practiced or worked more diligently than expected, you may see a firmness or soreness in your muscles. That solidness results from the development of lactic acid in the muscles. Presently eating fermented nourishments that contain lactic acid may not make you "hardened," yet does it appear to be insightful to eat foods that are as of now high in waste results.
- Ammonia is another result of fermentation. Vinegar, like acetic acid, likewise results from food fermentation.

2.6   FERMENTED FOODS—LOW IN NUTRITION

The foods that are most elevated in nourishment are those which are eaten in their new, characteristic, and natural state. When nourishment is messed with in any capacity, supplement misfortune results. The more drawn out food is held away, the lower it progresses toward becoming in nourishment. Another reason given for eating fermented nourishments is that they are high in B-vitamins, or that they may be one way or another urge the body to deliver more Vitamin B_{12} in its digestive organs. The polar opposite might be valid. As per look into, the degrees of Vitamin B_{12} might be decreased by fermented foods. Rather than adding dietary advantages to the nourishment, fermentation diminishes some nutrient and mineral accessibility. The nourishments that are most noteworthy in foods are those which are eaten in their crisp, common, and natural state. When food is messed with in any capacity, supplement misfortune results. The more drawn out nourishment is held away, the lower it progresses toward becoming in nutrition.

2.7   FERMENTATION STARTER CULTURES AND ITS BY-PRODUCTS

Starter cultures, which complete the fermentation procedure, are utilized to guarantee consistency in commercial products by utilizing known species with alluring attributes, for example, a high rate of fermentation through the generation of lactic acid as well as the emission of optional metabolites into the fermentate network (Figure 2.2). Novel starter societies are persistently sought after for the advancement of new advertisement products alongside more prominent characterization of those at present being used to guarantee sheltered and useful products [24–26].

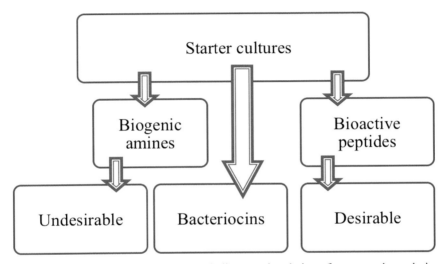

FIGURE 2.2   Impact of bioactive metabolites on the choice of starter culture during fermentation.

Fermentation starters can deliver various attractive and bothersome bioactive metabolites. Biogenic amines (BAs) (left) are bothersome products in much fermentation because of their lethality. Bioactive peptides (right) delivered through enzymatic discharge are alluring results because of positive natural movement. Bacteriocins (focus) are alluring as a known probiotic attribute, however conceivably bothersome in a starter culture because of conceivable effect on different fermenting cultures.

2.7.1   *BACTERIOCINS [27–34]*

Bacteriocins are little ribosomally orchestrated antimicrobial peptides against which the maker species is resistant and which act against other microbes in a bactericidal or bacteriostatic manner. Incredible consideration must be taken with respect to bacteriocin generation in starter societies, as they may target other fermenting cultures or aides; be that as it may, their capacity to repress potential deterioration microscopic organisms and pathogens can be of extraordinary use. The identification of bacteriocinogenic strains has, for the most part, depended on agar dissemination based assays. Expanding enthusiasm for bacteriocins as options in contrast to anti-toxins and synthetic foods additives has prompted new strategies for recognizing bacteriocin makers. *In silico,* screening utilizing projects, for example, BAGEL, and anti-SMASH enable the revelation of new bacteriocin operons where entire genome information is accessible. Such strategies keep away from any potential problems with unsusceptible pointer strains and can take into account quicker introductory screening. *In silico*, screens such as these depend on recently depicted peptides or the recognizable proof of bacteriocin accessory genes, and as such, it is unlikely that initial agar dispersion based measures will be totally supplanted with in silico screening. Be that as it may, they may speak to a chance to look for new bacteriocins in complex microbiotas, for example, those of conventional fermentation products.

2.7.2   *BIOGENIC AMINES (BAS) [35–45]*

Biogenic amines (BAs) are organically dynamic, low-sub-atomic weight natural bases delivered chiefly through the decarboxylation of certain amino acids, which can aggregate during fermentation. Generally, the nearness of BAs in nourishment products s is related with unfortunate microbial movement, demonstrating food deterioration or damaged manufacture. Dairy products, specifically cheese, can gather elevated amounts of BAs, predominantly histamine and tyramine, which are known to be toxic, yet starting at yet legitimate breaking points have been set distinctly for histamine in fish products. The collection of more than one sort of BA in items is of specific concern inferable from their synergistic poisonous quality at dietary focuses, which has as of late been exhibited with intestinal cells in vitro. BAs are recognized in dairy items by the chromatographic recognition of BA compounds or through the detection of BA-producer organisms utilizing PCR-based strategies, which relate with HPLC results. In an ongoing report, levels of tyramine in a model cheese were diminished by 85% using a bacteriophage to confine the populace of BA-producing bacteria. Pre-selection of starter cultures lacking BA qualities later on might be important to stay away from undesirable development of BA mixes and proceeded with shirking of defilement, which is known to happen during post-ripening processing.

2.7.3 BIOACTIVE PEPTIDES [46–50]

Bioactive peptides are encoded in bigger proteins and, when discharged after proteolysis, have been related with wellbeing advancement through various instruments, for example, restraint of angiotensin 1-changing over catalyst (ACE: angiotensin 1-converting enzyme) action, anti-thrombotic movement, antihypertensive action, cancer prevention agent action, immunomodulation, apoptosis modulation, and by opioid and anti-narcotic activities. LAB has a heap of proteases and peptidases that can discharge scrambled peptides during milk fermentation or following the ingestion of fermented products containing LAB in the intestine lumen. In recent years, potential anti-carcinogenic peptides have been found encoded in cow-like milk casein and whey proteins, including the recently known cationic lactoferricin. The known cancer-preventative peptide lunasin is proteolytically discharged during fermentation by LAB. Further research has in this way uncovered expanded protease obstruction during in vitro gastrointestinal travel within the sight of normally happening protease inhibitors to permit lunasin to reach the enormous intestine. Another ongoing investigation has discovered that the administration of milk products by a probiotic *Lactobacillus casei* strain modulated the immune response against a breast cancer tumor in a mouse model, with deferred or blocked tumor advancement in the fermented milk-encouraged gathering as contrasted and unfermented milk as a control. The system of activity has not yet been clarified, and investigations of a comparative sort have not yet passed the creature preliminary preclinical stages of the investigation.

2.7.4 ETHNIC FERMENTED MILK ITEMS

There is expanding enthusiasm for novel LAB strains secluded from ethnic fermented milk products, which would have been a piece of the autochthonous aging microbiota. Items, for example, matsoni, a fermented milk result of Armenian cause, and kulenaoto, the traditional fermented milk product of the Maasai in Kenya, are having their beforehand undescribed microbiot as portrayed utilizing sequencing-based analysis. Without a doubt, such items might be of extraordinary worth: for instance, shubat, a probiotic fermented camel milk of Kazakh starting point, has as of late been found to exhibit positive hypoglycemic action in sort diabetic rats, and the indigenous Indian fermented refreshment Raabadi has been explored as a wellspring of probiotic hypocholesterolemic lactobacilli. This is an unmistakable difference to the wellbeing cases made for most business fermented dairy products that show the advantages of utilizing protected, realized starter societies [51].

2.7.5 PREBIOTICS

In 2008, prebiotics were characterized by the International Scientific Association for Probiotics and Prebiotics (ISAPP) as "a specifically fermented fixing that outcomes in explicit changes in the composition as well as the action of the gastrointestinal microbiota, along these lines giving benefit(s) upon host health," a definition that is right now being additionally reexamined by ISAPP. Prebiotics are fermented by the gastrointestinal microbiota and add to solid tweak of the gut. The ingestion of explicit prebiotics has been appeared to expand the antibacterial capacities of a probiotic strain. Synbiotics are a generally new territory that includes a mix of probiotic and prebiotic in one product; the prebiotic is planned to improve the survival/ development/ performance of the probiotic or other beneficial microscopic organisms in the colon, which thus has valuable wellbeing impacts on the host. As of now, inquire about is being led on the job of the gut microbiota in the advancement of malignancy, with an attention on colorectal cancer. While this exploration is in its beginning period, there is proof for the utilization of probiotics, prebiotics, and synbiotics in the treatment or counteractive action of this ailment. There is potential for these to go about as anti-carcinogens or anti-mutagenic specialists through eating regimen-based intercessions. Progressively point by point examination could prompt immense walks in the anticipation of malignancy; however starting at yet the field is available to new research [52, 53].

2.7.6 WELLBEING CENTERED RESEARCH

An investigation into the utilization of fermented foods as a potential way to deal with the battle ailment is developing, yet it must be valued that a large number of these practical nourishments are proposed to counteract infection beginning, or alleviate side effects, and not really go about as a therapeutic agent. This builds the weight of confirmation on the analyst to demonstrate that the fermentation of the prebiotic was for sure the reason the host stayed solid. Regulation of the gut microbiota is the focal point of numerous examinations relating to microbial fermenting to measurable health benefits. One rising zone of study utilizing microbial fermentation is in osteoporosis. Osteoporosis is normal in post-menopausal ladies and the old and presents itself as debilitated bones inclined to breaks or cracks because of poor calcium absorption. The utilization of a prebiotic fructooligosaccharide (FOS) can possibly be a safeguard strategy for osteoporosis. The prebiotic is fermented in the gut, causing a drop in the pH of the lumen to such a degree, that already insoluble calcium phosphate will break down. This assumes a helpful job in bone mineral density. The fermentation of FOS releases short-chain fatty acids and lactic acid, which cause a drop in pH. This broke up calcium brings about an expansion in inactive dissemination; therefore, it could treat, or even possibly avoid, the beginning of osteoporosis.

Obesity is a global issue and has generated much interest in whether and how our gut microscopic organisms could be a contributing variable in the improvement of this unpredictable disorder. On-going examinations into the gut microbiota are aimed at identifying whether a specific bacterium or bacterial gathering could be adding to obesity. While this is a rising region of research, there are energizing improvements on the best way to conceivably battle this disorder through the regulation of the gut microbiota. The purported "fat microbiota profile" can be described as a diminished bacteroidetes/firmicutes proportion in individuals. One investigation took a gander at the organization of prebiotics, for example, FOS as a potential strategy for diminishing the probability of weight by expanding the degrees of "lean microbiota" through fermentation of the prebiotic in the gut. This thusly prompted a lessening in the porousness of the digestive system with improved tight intersection trustworthiness. While this is in beginning times of research, it presents a conceivably new technique by which heftiness could be treated through microbial fermentation inside the gut. This region of research has extraordinary potential medicinally and commercially [54, 55].

2.7.7   NATIONAL PROPOSALS

Fermented foods have been expended worldwide for a great many years before any immediate medical advantages were genuinely comprehended. While the interest from purchasers for utilitarian foods is developing, the national suggestions are not going with the same pattern. Presently that the systems by which these fermentations can helpfully affect human health are beginning to be elucidated, food guidelines around the globe are gradually starting to suggest their utilization. This consideration isn't all-inclusive, notwithstanding verifiable use and clinical preliminaries demonstrating the advantages of these fermented products in the eating regimen. Given the solid custom of fermented nourishments in Asia, it is to some degree astonishing that they are not explicitly incorporated into food rules, yet the Chinese Nutrition Society recommends the consumption of yogurt for the individuals who don't endure lactose well.

There is a high occurrence of lactose narrow mindedness in Asian nations, and there are clinically demonstrated investigations that demonstrate the incorporation of fermented dairy nourishments can ease the indications of intolerance. Japanese experts rundown fermented foods in the food of specified health use (FOSHU) class, and in India the rules explicitly energize the utilization of fermented nourishments. The Indian guide features explicitly that pregnant ladies should consider including more fermented foods in their diet inferable from the expanded bioavailability of iron that is related with these foods.

2.7.8   GUIDELINE OF FERMENTED DAIRY PRODUCTS

The guideline encompassing microbial fermentation in the nourishment industry is beginning to have a detrimental effect on the industry as an entirety. For instance, there is as of now no lawful definition for the term "probiotics"; until scientific, legal, and industrial teams are all cooperating under one strong definition, the expression "probiotics" will begin to lose its meaning. Along with this, the general communities are losing confidence in the benefits that offer fermented dairy products that are enhanced with probiotic/prebiotics. This change has prompted shopper disarray as to whether or not the claims were ever true. It is essential that labeling of fermented food products with clinically demonstrated medical advantages is allowed to enable the industry to start to benefit from subsidizing these trials, or they will begin to invest in marketing strategies rather than the genuinely necessary research. The International Dairy Federation (IDF) represents the dairy sector at relevant codex meetings regarding the global guidelines for dairy products. The IDF is presently associated with exploring product naming as to wholesome data and wellbeing cases and how these influence the shopper's decision of various products. These examinations will trustfully lead to a change in labeling laws to allow for clinically proven wellbeing professes to be available on fermented dairy products.

2.8 CURRENT MARKET VALUE

The strategies for fermenting have been gone on through generations utilizing locally accessible crude materials from plants and creatures. Individuals around the globe produce fermented nourishment and drink overflowing with smaller scale life forms, either normally or by including a starter culture Takamine, in 1884, was conceded US patent (Nos. 525, 823) for a procedure for making diastatic compound from the koji shape, which was in all respects effectively advertised as Taka-diastase for commercial use. According to Tamiya [5], food and drinks produced with koji retailed for $1,000 million per year in Japan, which progressed toward becoming $205,000 million out of 1970. Beginning in about the 1960s and expanding quickly after the mid-1970s, East Asian matured soy foods (particularly soy sauce or shoyu, miso, and tempeh, in a specific order), started to be generally utilized in the West.

Research Moz Latest Publication on Fermentation Chemicals Market-Global Industry Analysis, Size, Share, Growth, Trends, and Forecast, 2013–2019 made an investigation, which is done dependent on volumes (kilo tons) and income (USD million) for the period going from 2013 to 2019. The examination incorporates the drivers and limitations for the maturation synthetic substances showcase alongside the effect and the investigation of the open doors accessible in the aging synthetic substances advertise on a worldwide and provincial level. The worldwide market for mechanical chemicals was esteemed at $3.1 billion of every 2009 and came to about$3.6 billion in 2010 [56]. The estimated market for 2011 is about $3.9 billion. BCC ventures this market to develop at an aggravated yearly development rate (CAGR) of 9.1% to reach $6 billion by 2016. Nourishment and refreshment chemicals include the biggest section of the mechanical compounds industry with incomes of about $1.2 billion out of 2010. This market is relied upon to reach $1.3 billion by 2011 and further, it will develop to $2.1 billion by 2016, a CAGR of 10.4%. The second-biggest class is specialized compounds with incomes of about $1.1 billion in 2010 and nearly $1.2 billion in 2011. This market is further expected to develop to $1.7 billion by 2016, a CAGR of 8.2%. Mechanical catalyst applications have the second-biggest portion of the market and are relied upon to develop at a CAGR of 8.9%, from $3.2 billion of every 2008 to an expected $4.9 billion in 2013. The worldwide market for aging items was required to reach $22.4 billion before the finish of 2013, a compound yearly development rate (CAGR) of 7.0%, Amino acids have the biggest portion of the market, creating $7.8 billion in 2013. The complete estimation of the world aging generation will appear to be very little unique than today as far as to limit or as far as item yield or incentive in not so distant future. Much development can't be normal. Consequently, the entire maturation industry is confronting a hazy future. As far as worth, amino acids, natural acids, and catalysts, just as anti-infection agents, are of around equivalent size with about $1 billion each. As far as to limit (and this fills in as a parameter for the creation per unit of fermenter size), anti-infection agents lead the pack with 40% of the worldwide limits of around 360,000 to 370,000 m^3.

The vast majority of the natural and amino acids are utilized in feed or nourishment, and most anti-toxins in human treatment and in the feed. Advancements in the feed showcase and somewhat in the food market have, in this manner, exceptional significance for the fate of such products. The feed market is experiencing scaled-down interest for meat, and lower meat costs also. In Europe, meat utilization is diminishing a direct result of the wellbeing angle, and in Eastern Europe and Asia, on account of the decreased buying intensity of the populace. Nutrients and polymers, the littler aging items, are no special cases. In these fields, value decreases will likewise be the main impetus for a propensity toward lower market an incentive in a couple of years. Conceivable higher interest in volume terms won't most likely redress [56].

2.9   PROSPECT OF FERMENTED FOOD

Aside from chemicals, sugars, alcohols, nutrients, amino acids, anti-infection agents, the aging item itself can be utilized as a food supplement. Probiotics are generally characterized as microbial food supplements with useful impacts on the shoppers. Élie Metchnikoff is viewed as the designer of probiotics captivated by the life span of the Caucasian populace and its successive utilization of matured milks, recommended that the corrosive creating life forms in fermented dairy products could counteract fouling in the digestive organ and in this way lead to a prolongation of the life expectancy of the customer. Proof based surveys show that specific strains of probiotics add to the microbial parity of the gastrointestinal tract (GI tract) and wellbeing conditions that can profit by probiotics treatment incorporate looseness of the bowels, gastroenteritis, touchy inside disorder, provocative gut malady, and cancer. The most normally utilized living beings in probiotic arrangements are the lactic corrosive microscopic organisms.

Be that as it may, precisely which probiotic strains, proper measurements, and fermenting profiles are as yet being researched. As per the paper "Probiotics and Prebiotics in Dietetics Practice" in the March 2008 *Journal of the Academy of Nutrition and Dietetics*, the test in creating clinical suggestions for probiotics treatment is an absence of combined research and consistency crosswise over examinations regarding bacterial strains, measurements, and populaces [59].

Notwithstanding supporting human wellbeing, Lactobacillus, and other microscopic organisms may secure against foodborne ailment by restraining and killing foodborne pathogens, including Listeria monocytogenes, *Staphylococcus aureus*, and *Bacillus cereus*. The restraint of pathogenic microscopic organisms might be expected to a limited extent to pH, just as antimicrobial bacteriocins delivered by Lactobacillus to hinder other focused strains, including foodborne pathogens. While these discoveries bolster aging as a sheltered strategy for protection, and utilization of matured grain has been related with abatements in foodborne ailment, more research is required. Quality innovation will absolutely assume a job in growing new strains, with quality sequencing taking into consideration an expanded comprehension of instruments and usefulness of probiotics. One procedure utilized by the integrative drug is the reintroduction of gainful microscopic organisms to improve stomach related capacity and rebalance the intestinal verdure, which is cultivated by the probiotics. After some time, new sustenance items containing probiotics will develop, for example, vitality bars, oats, juices, baby equation, and cheddar, just as ailment explicit medicinal nourishments. The foundation of benchmarks of character for probiotic-containing sustenance items will serve to quicken the improvement and accessibility of these nourishment products. While probiotic supplementation is widely utilized, many prefer using a food first approach by suggesting normally fermented foods [57].

2.10   CURRENT RESEARCH TRENDS IN MODERN FOOD FERMENTATION

Current biotechnology empowers plants, animals, and microorganisms to be hereditarily adjusted (GM) with the incorporation of required novel characteristics. The consideration of the ideal novel attributes conceivably offers improved quality, nutrition, and processing attributes, which can permit the creation of progressively nutritious, more secure, more delicious, and more advantageous nourishment. Procedures in biotechnology, especially propel in recombinant innovation have been changing our customary method for nourishment creation and planning by improving the foods we eat and the refreshments we drink a long way past the conventional degree. Presently, the change has concentrated more on characteristics which improve, healthful quality, tangible quality, bacteriophage opposition, capacity to create antimicrobial mixes, for example, bacteriocins, corruption or inactivation of common poisons, for example, cyanogenic glucosides, mycotoxins, hostile to wholesome factors, for example, phytates, end of cancer-causing mixes, for example, ethyl carbamate in mixed drinks. Another uncommon spotlight is on chemicals, the natural impetuses used to encourage and accelerate metabolic responses in living life forms. Some progressed biotechnological look into models is featured.

2.10.1   CYANOGENIC GLYCOSIDES IN CASSAVA

Cassava (*Manihot esculenta Crantz*) is broadly utilized all through Africa as a significant food source yet its serious issue is that it contains harmful mixes called cyanogenic glycosides, linamarin, and lotaustralin which free cyanide. Nonetheless, these conventional practices are overwhelmed with so many issues for which current biotechnology offers the best arrangement. Biotechnological look into on cassava aging created proper starter culture that can deliver amylase and linamarase compounds fundamental for starch breakdown and cyanogenic glucoside hydrolysis, separately. Strains of Lactobacillus plantarum were found to create these catalysts. As indicated by Oyewole, to start hereditary control of cassava lactic corrosive microorganisms, the plasmid profiles of the lactobacilli disengaged from cassava were examined and the nearness of plasmids has been affirmed and further biotechnological research approach has been given trust in exploring improvement of the procedure.

2.10.2   ETHYL CARBAMATE IN ALCOHOLIC BEVERAGES

Ethyl carbamate or urethane is one of the bothersome results shaped during yeast fermentation of nourishments and refreshments and it is a potentially cancer-causing substance. Recombinant DNA innovations has made conceivable to bring new properties into the yeast, just as to wipe out the undesirable by-items. Ethyl carbamate is orchestrated by the unconstrained response among ethanol and urea. Yeasts utilized in wine maturation have arginase catalyst that catalyzes debasement of arginine (found in the huge sum in grapes) into urea.

2.10.3   ACRYLAMIDE IN FOODS

Acrylamide is a suspected cancer-causing and harmful substance, whose nearness in nourishments was first distinguished in 2002. A wide range of cooked foods contains acrylamide at the various sum. Anese et al. has summarized systems of acrylamide arrangement and the most significant measures in regards to acrylamide alleviation including sustenance maturation. Be that as it may, utilization of biotechnological devices is one and considered generally encouraging. Because of this biotechnological devices, chemical asparaginase from cloned *Aspergillus oryzae* has been created and is professed to diminish acrylamide levels by up to 90% through the change of asparagine into aspartic corrosive without modifying the appearance or taste of the last product.

2.10.4   ENZYMES, AMINO ACIDS, NUTRIENTS, BIOFLAVORS, AND CAROTENOIDS

These days, numerous compounds utilized in sustenance handling businesses are inferred using hereditarily changed microorganisms (GMMs). Hereditarily changed yeasts, parasites, and microorganisms have been in business use for this reason for over 10 years. Models include: chymosin for cheddar making. Chymosin, usually known as rennin, is the chief milk-coagulating chemical present in rennet. As of now, recombinant DNA innovation has made it conceivable to acquire chymosin from strains of microorganisms.

Hereditarily altered microorganisms have likewise been being used for the creation of different substances, for example, micronutrients (nutrients, amino acids), bioflavors, and carotenoids. A large portion of the microorganisms changed for sustenance handling are subsidiaries of microorganisms utilized in customary nourishment biotechnology. Microorganisms improved by present-day biotechnology are likewise a work in progress in the field of probiotics. These days, probiotic foods get market enthusiasm as wellbeing advancing practical nourishments.

2.10.5   RESEARCH ON ETHIOPIAN INDIGENOUS TRADITIONAL FERMENTED FOODS AND BEVERAGES

Ethiopia is supplied with ethnic assorted variety showed in social decent variety and an assortment of customary aged sustenance and refreshments. As indicated by Mogessie, a microbiological study on Ethiopian conventional fermented foods and beverages has begun during the 1980s on points like microbial progression and going with changes, sanitation, preparing, and waste. The vast majority of these works were focused on some in all respects usually and broadly real-ized conventional fermented foods and beverages about around the focal and southern pieces of the nation.

Be that as it may, in light of worldwide advances in food fermenta-tion biotechnology and commercialization/industrialization, inquire about on Ethiopian indigenous conventional fermented foods and beverages are not yet drawing closer to the desires [57].

2.11   CONCLUSION

Sometime before probiotics turned into the known similarly as with restorative worth, societies since the beginning and over the globe were celebrating fermented nourishments. The use of valuable genomics and metabolomics to microbial cell processing plants will surely enhance the fermenting innovation and open new roads for the savvy and eco-friendly generation of significant worth included results of human intrigue. Microbial fermentation holds the way to some incredibly intricate connections between bacterial species and the nourishment grid they are fermenting. The examinations featured in this audit demonstrate the capability of using this microbial fermentation in a more learning-based style than that of the past. As to microbial fermenting in nourishment, this speaks to a region with potential well past the expansion of time span of usability. The work in these regions is proceeding, with the assistance of better guidelines, could prompt energizing new disclosures on overseeing ailment side effects through foods. There have been various advances in fermented products. With the general conclusion moving towards more advantageous ways of life and survey substance additives in a negative light, fermented products show an extraordinary business guarantee. New starter societies are being distinguished utilizing progressively modern techniques to guarantee their adequacy and suitability. In silico-based techniques and research in the wellbeing, advancing exercises of LAB in fermentates are on the ascent, alongside the characterization of conventional products that have been related with great wellbeing.

KEYWORDS

- biogenic amines
- lactic corrosive microbes
- microbial fermentation
- microorganisms
- strong state fermentation
- submerged fermentation

REFERENCES

1. McGovern, P. E., Zhang, J., Tang, J., Zhang, Z., Hall, G. R., Moreau, R. A., Nuñez, A., et al., (2004). Fermented beverages of pre-and proto-historic China. *Proc. Natl. Acad. Sci. U.S.A, 101*(51), 17593–17598.
2. Dirar, H. *The Indigenous Fermented Foods of Sudan: A Study in African Food and Nutrition.* Oxford University Press, UK.
3. Sahrhage, D., (2008). Fishing in the stone age. *Encyclopedia of the History of Science, Technology, and Medicine in Non-Western Cultures*, pp. 935–939.
4. Shurtleff, W., & Aoyagi, A. *A Brief History of Fermentation.* East and West; A chapter from the manuscript, History of Soybeans and Soy Foods, 1100 B.C. to the 1980s.
5. Tamiya, H., (1958). The Koji, an important source on enzymes in Japan. *Proc. Int. Symp. Enz. Chem.* Tokyo and Kyoto.
6. Talkad, M. S., Acharya, S., Das, M. J., Sharma, N., & Shantha, S. L., (2012). Comparative study of lipase and protease enzymes in two different sources. *Int. J. Adv. Biotechnol. Res., 3*, 814–823.
7. Hesseltine, C. W., Smith, M., & Wang, H. L., (1967). New fermented cereal products. *Devel. Ind. Microbiol.,* pp. 8179–8186.
8. Subramaniyam, R., & Vimala, R., (2012). Solid-state and fermentation for the production of bioactive substances: A comparative study. *Int. J. Sci. Nat., 3*(3), 480–486.
9. Martins, D. A. B., Do Prado, H. F. A., Leite, R. S. R., Ferreira, H., De Souza, M. M. M., Da Silva, R., & Gomes, E., (2011). Agro-industrial wastes as substrates for microbial enzyme production and source of sugar for bioethanol production. In: Sunil, K., (ed.), *Integrated Waste Management* (Vol. II). In Tech.
10. Ejtahed, H. S., Soroush, A. R., Angoorani, P., et al., (2016). Gut microbiota as a target in the pathogenesis of metabolic disorders: A new approach to novel therapeutic agents. *Horm. Metab. Res., 48*(6), 349–358.
11. Martinez, R. C., Bedani, R., & Saad, S. M., (2015). Scientific evidence for health effects attributed to the consumption of probiotics and prebiotics: An update for current perspectives and future challenges. *Br. J. Nutr., 114*(12), 1993–2015.
12. Parvez, S., Malik, K. A., Ah-Kang, S., et al., (2006). Probiotics and their fermented food products are beneficial for health. *J. Appl. Microbiol., 100*(6), 1171–1185.
13. Owusu-Kwarteng, J., Tano-Debrah, K., Akabanda, F., et al., (2015). Technological properties and probiotic potential of *Lactobacillus fermentum* strain isolated from West African fermented millet dough. *BMC Microbiol., 15*, 261.
14. Ao, X., Zhang, X., Zhang, X., et al., (2012). Identification of lactic acid bacteria in traditional fermented yak milk and evaluation of their application in fermented milk products. *J. Dairy Sci., 95*(3), 1073–1084.
15. Anandharaj, M., Sivasankari, B., Santhanakaruppu, R., et al., (2015). Determining the probiotic potential of cholesterol-reducing *Lactobacillus* and *Weissella* strains isolated from gherkins (fermented cucumber) and south Indian fermented koozh. *Res. Microbiol., 166*(5), 428–439.

16. Aneja, R. P., Vyas, M. N., Nanda, K., & Thareja, V. K., (1977). Development of an industrial process for the manufacture of Shrikhand. *J. Food Sci. Technology.*
17. Kanawjia, S. K., (2006). Developments in the manufacture of lassi. In: *Developments in Traditional Dairy Products.*
18. James, M. J., (2000). *Fermentation and Fermented Dairy Products.*
19. Lathasabikhi, (2006). *Developments in Traditional Dairy Products.*
20. Ryan, P. M., Ross, R. P., Fitzgerald, G. F., et al., (2015). Sugar-coated: Exopolysaccharide producing lactic acid bacteria for food and human health applications. *Food Funct., 6*(3), 679–693.
21. Pessione, E., & Cirrincione, S., (2016). Bioactive molecules released in food by lactic acid bacteria: Encrypted peptides and biogenic amines. *Front Microbiol., 7*, 876.
22. Dobson, A., Cotter, P. D., Ross, R. P., et al., (2012). Bacteriocin production: A probiotic trait? *Appl. Environ. Microbiol., 78*(1), 1–6.
23. Hegarty, J. W., Guinane, C. M., Ross, R. P., et al., (2016). Bacteriocin production: A relatively unharnessed probiotic trait? [Version 1; Referees: 2 approved]. *F1000 Res., 5*, 2587.
24. Galvez, A., Lopez, R. L., Abriouel, H., et al., (2008). Application of bacteriocins in the control of foodborne pathogenic and spoilage bacteria. *Crit. Rev. Biotechnol., 28*(2), 125–152.
25. Wolf, C. E., & Gibbons, W. R., (1996). An improved method for quantification of the Bacteriocin nisin. *J. Appl. Bacteriol., 80*(4), 453–457.
26. Van, H. A. J., De Jong, A., Montalbán-López, M., et al., (2013). BAGEL3: Automated identification of genes encoding bacteriocins and (non-)bactericidal post-translationally modified peptides. *Nucleic Acids Res., 41*(Web server issue), W448–453.
27. Blin, K., Medema, M. H., Kottmann, R., et al., (2017). The anti-SMASH database, a comprehensive database of microbial secondary metabolite biosynthetic gene clusters. *Nucleic Acids Res., 45*(D1), D555–D559.
28. Papagianni, M., Avramidis, N., Filioussis, G., et al., (2006). Determination of bacteriocin activity with bioassays carried out on solid and liquid substrates: Assessing the factor "indicator microorganism." *Microb. Cell. Fact., 5*, 30.
29. Collins, F., O'Connor, P. M., Gómez-Sala, B., et al., (2017). Bacteriocin gene-trait matching across the complete lactobacillus pangenome. *Sci. Rep.* Awaiting print.
30. Walsh, C. J., Guinane, C. M., Hill, C., et al., (2015). *In silico* identification of bacteriocin gene clusters in the gastrointestinal tract, based on the human microbiome project's reference genome database. *BMC Microbiol., 15*, 183.
31. Linares, D. M., Martin, M. C., Ladero, V., et al., (2011). Biogenic amines in dairy products. *Crit. Rev. Food Sci. Nutr., 51*(7), 691–703.
32. Linares, D. M., Del Rio, B., Redruello, B., et al., (2016). Comparative analysis of the *in vitro* cytotoxicity of the dietary biogenic amines tyramine and histamine. *Food Chem., 197*(Pt A), 658–663.
33. EFSA Panel on Biological Hazards (BIOHAZ), (2011). Scientific Opinion on risk-based control of biogenic amine formation in fermented foods. *EFSA Journal, 9*(10), 2393.

34. Ladero, V., Cañedo, E., Pérez, M., et al., (2012). Multiplex qPCR for the detection and quantification of putrescine-producing lactic acid bacteria in dairy products. *Food Control., 27*(2), 307–313.

35. Regulation EC, (2005). 2073/2005 of 15 November 2005 on microbiological criteria for foodstuffs. *Official Journal of the European Union, 338*, 1–29.

36. Del Rio, B., Redruello, B., Linares, D. M., et al., (2017). The dietary biogenic amines tyramine and histamine show synergistic toxicity towards intestinal cells in culture. *Food Chem., 218*, 249–255.

37. Fernández, M., Linares, D. M., Del Rio, B., et al., (2007). HPLC quantification of biogenic amines in cheeses: Correlation with PCR-detection of tyramine-producing microorganisms. *J. Dairy Res., 74*(3), 276–282.

38. Ladero, V., Gómez-Sordo, C., Sánchez-Llana, E., et al., (2016). Q69 (an E. faecalis-infecting bacteriophage) as a biocontrol agent for reducing tyramine in dairy products. *Front Microbiol., 7*, 445.

39. Ladero, V., Fernández, M., & Alvarez, M. A., (2009). Effect of post-ripening processing on the histamine and histamine-producing bacteria contents of different cheeses. *Int. Dairy J., 19*(12), 759–762.

40. Diaz, M., Del Rio, B., Sanchez-Llana, E., et al., (2016). Histamine-producing *Lactobacillus parabuchneri* strains isolated from grated cheese can form biofilms on stainless steel. *Food Microbiol., 59*, 85–91.

41. Hayes, M., Ross, R. P., Fitzgerald, G. F., et al., (2007). Putting microbes to work: Dairy fermentation, cell factories, and bioactive peptides. Part I: Overview. *Biotechnol. J., 2*(4), 426–434.

42. Savijoki, K., Ingmer, H., & Varmanen, P., (2006). Proteolytic systems of lactic acid bacteria. *Appl. Microbiol. Biotechnol., 71*(4), 394–406.

43. Pepe, G., Tenore, G. C., Mastrocinque, R., et al., (2013). Potential anti-carcinogenic peptides from bovine milk. *J Amino Acids.,* p. 939804.

44. Rizzello, C. G., Nionelli, L., Coda, R., et al., (2012). Synthesis of the cancer-preventive peptide lunasin by lactic acid bacteria during sourdough fermentation. *Nutr.Cancer., 64*(1), 111–120.

45. Cruz-Huerta, E., Fernández-Tomé, S., Arques, M. C., et al., (2015). The protective role of the bowman-birk protease inhibitor in soybean lunasin digestion: The effect of released peptides on colon cancer growth. *Food Funct., 6*(8), 2626–2635.

46. Aragón, F., Carino, S., Perdigón, G., et al., (2014). The administration of milk fermented by the probiotic *Lactobacillus casei* CRL 431 exerts an immunomodulatory effect against a breast tumor in a mouse model. *Immunobiology., 219*(6), 457–464.

47. Bokulich, N. A., Amiranashvili, L., Chitchyan, K., et al., (2015). Microbial biogeography of the transnational fermented milk matsoni. *Food Microbiol., 50*, 12–19.

48. Hill, C., Guarner, F., Reid, G., et al., (2014). Expert consensus document. The International Scientific Association for probiotics and prebiotics consensus statement on the scope and appropriate use of the term probiotic. *Nat. Rev. Gastro. Enterol. Hepatol., 11*(8), 506–514.

49. Di Bartolomeo, F., Startek, J. B., & Van, D. E. W., (2013). Prebiotics to fight diseases: Reality or fiction? *Phytother. Res., 27*(10), 1457–1473.

50. Compare, D., Rocco, A., SanduzziZamparelli, M., et al., (2016). The gut bacteria-driven obesity development. *Dig Dis., 34*(3), 221–229.

51. Madjd, A., Taylor, M. A., Mousavi, N., et al., (2016). Comparison of the effect of daily consumption of probiotic compared with low-fat conventional yogurt on weight loss in healthy obese women following an energy-restricted diet: A randomized controlled trial. *Am. J. Clin. Nutr., 103*(2), 323–329.

52. BCC Research, (2012). *Global Markets for Enzymes in Industrial Applications.*

53. Gordon, S., (2008). Elie metchnikoff: Father of natural immunity. *Eur. J. Immunol., 38*, 3257–3264.

54. Anon, (1992). Research priorities in traditional fermented foods. In: *Application of Biotechnology to Traditional Fermented Foods* (pp. 3–8, 199). Report of an Ad Hoc Panel of the Board on Science and Technology for International Development, National Academy Press Washington, D.C.

55. Oyewole, O. B., (1992). Cassava processing in Africa. In: *Application of Biotechnology to Traditional Fermented Foods* (pp. 89–92). Report of an Ad-Hoc Panel of the Board on Science and Technology for International Development, National Academy Press. Washington D.C.

56. Joana, C., John, I. H., Debra, L. I., George, K., Van, D. M., Aline, L., Daniel, J. E., & Hennie, J. J. V. V., (2006). Metabolic engineering of *Saccharomyces cerevisiae* to minimize the production of ethyl carbamate in wine. *Am. J. Enol. Vitic., 57*(2), 113–124.

57. Council for Agricultural Science and Technology (CAST), (2006). *Acrylamide in Foods.* Issue Paper No. 32.

58. Govinden, R., Pillay, B., Van, Z. W. H., & Pillay, D., (2001). Xylitol production by recombinant *Saccharomyces cerevisiae* expressing the pichiastipitis and Candida shehatae XYL1 genes. *Appl. Microbiol. Biotechnol., 55*, 76–80.

59. Douglas, L. C., Sanders, M. E., (2008). Probiotics and prebiotics in dietetics practice. J Am Diet Assoc., *108*(3), 510–21.

CHAPTER 3

Recent Reports on Imperative Medicinal Potentials of *Boswellia serrata*

SOURANAVA JANA,[1] DEBARSHI KAR MAHAPATRA,[2] and SOUVIK MUKHERJEE[1]

[1]*Institute of Pharmaceutical Sciences, Guru Ghasidas Viswavidyalaya (A Central University), Bilaspur – 495009, Chhattisgarh, India, E-mail: mukherjees388@gmail.com (S. Mukherjee)*

[2]*Department of Pharmaceutical Chemistry, Dadasaheb Balpande College of Pharmacy, Nagpur – 440037, Maharashtra, India*

ABSTRACT

Boswellia serrata is thought of as an Ayurvedic medication and intimate Burseraceae class. The plant is cosmopolitan in India and grows in the dry steep forests of provinces such as Rajasthan, Madhya Pradesh, Gujarat, Bihar, Assam, Orissa, etc. Hindus, Babylonians, Persians, Romans, Chinese, Greeks, and Yankee civilizations used it primarily for ritual, embalming, and for its incense in cultural functions. This plant is mentioned in ancient Unani texts as an efficient remedy for respiratory disease, asthma, cough, vessel diseases, diarrhea, dysentery, ringworm, boils, fevers (antipyretic), skin, and blood diseases, mouth sores, channel discharges, etc. The qualitative phytochemical examination of plant extract indicates the presence of tannic acid, pentosans, lignin, holocellulose, β-sitosterol, volatile (cadinene, eleneol, gereniol, linalool, β-pinene, phenols, terpenyl acetate, bornyl acetate, etc.) and non-volatile (diterpene alcohol serratol, α-amyrin, and β-amyrin, triterpenic acids *viz.*, boswellic acid) oils. It primarily possesses anti-arthritic, medicine, anti-hyperlipidemic, anti-cancer, hypoglycemic, anti-asthmatic, analgesic, hepatoprotective, etc. pharmacological activities.

3.1 INTRODUCTION

Human has been victimization herbs merchandise known as a healthful plant for combating disorders since the ancient time. The incense tree comes within the family of rosid dicot family that concerning seventeen genera and 600 species. Boswellia tree is a tropical shrub found in northeast Africa, Arabia, and tropical America, South Asia, etc. [1]. The tree found on dry hills and slopes, on gravelly soils between associate in Nursing altitude up to 1150 m, with annual temperature 0–45°C and mean annual downfall concerning 500–2000 metric linear unit. For correct growing of Boswellia tree, the soil with characteristics of rocky ridges. The species has the flexibility to grow within the poorest and therefore the shallowest soils [2]. In India, a lot of unremarkably found in thirsty forests of Madhya Pradesh, Gujarat, Bihar, Punjab, Chhattisgarh, province, and a few elements of the Western mountain range [3, 4]. It is commonly known in India in various regional languages in different names like Kundur, Salai, Luben (Hindi); Kundur (Bengali); Dhup, Gugali (Gujarati); Chitta, Guguladhuph (Kanada); Parangi, Saambraani (Tamil, Telugu, and Malayalam) and Ashvamutri, Kundara, Shallaki (Sanskrit) [5]; and Ru Xiang (Chinese) [6].

Several celebrated species apart incense tree embrace, genus *Boswellia sacra*, genus *Boswellia carterii*, genus *Boswellia papyrifera*, genus *Boswellia neglecta*, genus *Boswellia frerana*, *Boswellia ovalifoliolata*, and genus *Boswellia rivae*, etc. that square measure necessary in healthful purpose [6]. The organic compounds made manufacturing by *B. serrata* is additionally called Indian gum olibanum and *Boswellia serrata* guggul, found in dehydrated mountain sections of India [5]. The word "Frankincense" that derived from French that means authentic, true, or pure incense wherever another word "Olibanum" that derived from the Arabic language that means milk [7]. The taxonomical feature of the Boswellia plant is given in Figure 3.1.

Salai-guggal is a dried oleo gum resin obtained from exudates on the injury from the bark of the *B. serrata* tree. The gum natural resin extract of *B. serrata* having varied medical specialty action with potent medicinal drug properties that historically recycled as a people drugs to delight chronic inflammatory disorders. The tree is deciduous, a balsamiferous tree with a lightweight, spreading crown, and somewhat drooping branches. It always incorporates a short bole, length in concerning 3–5 m usually; it has a girth of 1–1.8 m and a height of 18–25 m [8]. Trees become leafless during the entire period of flowering and fruiting (Figure 3.2). The bark is incredibly skinny; the outer layer is greyish-green color, ruby-red, or ash-grey with a chlorophyll layer below the skinny outer layer that peels off being papyraceous with a moderate-sized to massive. Leaves alternate, exstipulate, odd-pinnate, 20–45 cm long, swarming towards the ends of the branches; leaflets concerning 17–31 cm, opposite, 2.5–8 cm × 0.8–1.5 cm, basal pairs usually smallest, sessile, lanceolate, crenate, ovate-lanceolate variable in size. Flowers are white, in stout racemes panicles, 10–20 cm long, shorter than the leaves, ends of the branches area unit crowded, however, there is no terminal roll persistent, pubescent outside, five to 7-toothed; teeth tiny, deltoid. Petals 5–7 erect, free, 0.5 cm long. Fruits or drupes are 3 cm long, trigonous, with 3 valves and 3 cordate, 1-seeded pyrenes, and winged at the side of the margins. Fruits are grown in March-April and fruits within the winter [2].

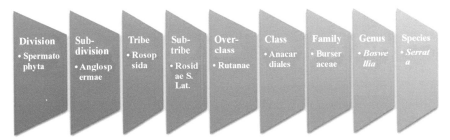

FIGURE 3.1   Taxonomical classification of *Boswellia serrata*.

The inflorescence could be a terminal flower cluster and produces up to ninety bisexual, actinomorphic flowers. On the average total spore, grain production is 10044 ± 1259 which concerns with the eighty-fifth of the contemporary spore grains area unit viable. The spore to ovule magnitude relation is 3348:1. The stigma is of the wet appendage kind. The fashion is hollow with 3 planar stylar canals crammed with a secretion product. A layer of organ canal cells borders the stylar canals. The ovary is trilocular and containing 3 ovules, one in every bodily cavity [9]. The rubber rosin taken from the rosid dicot genus tree hardens into lumps of varied shapes. They are yellow to chromatic in color on an irregular basis spherical or like stalactites, light-weight gray with a dusting, conjointly straightforward to interrupt with a conchoidal, waxy broken surface.

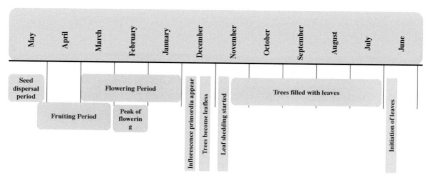

FIGURE 3.2   Various annual botanical events in *Boswellia serrata*.

The tree on incision exudes an oleo-gum organic compound by sound at the peak of fifteen cm from the bottom of the tree. The gathering of organic compound done at weekly intervals up to night and trees are often tapped for 6 to 8 years. The sound begins within the month of November and continues before the monsoon [10]. The oleo-gum organic compound collected in a very circular receptacle once scrapped off. Vegetable impurities area unit manually removed and unbroken in a very basket for a month on a cemented slopping floor to separate oil half. The remaining solid to semi-solid half referred to as gum-resin with traces of oil. Natural resin subjected to dry and typically treated with a talc ensuing crispness in the product. Further, it is broken into items and hierarchic picket mallet or chopper [3]. Every tree will offer concerning 1–1.5 kg of natural resin annually [2]. Four different kinds of qualities of natural resin (Figure 3.3) found in the market in step with its flavor, color, form, and size [5].

Macroscopically, the gum-resin happens to be clear, yellow-brown, varied shapes, and sizes up to 5 cm long, 2 cm thick, and stalactitic. The natural resin is odor with pleasant on burns with a transparent steady flame, it becomes with agreeable odor, style aromatic and agreeable, fracture brittle, fractured surface waxy, and clear. Microscopically, the rubble of fibers, rectangular cork cells, only a few yellow oil globules and various little or giant, oval to spherical or parallelogram crystalline fragments gift [5, 8]. The pharmacopeia limits for the various parameters of resin are given in Table 3.1.

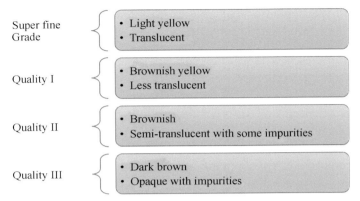

FIGURE 3.3   Quality of gum resin obtained from *Boswellia serrata*.

TABLE 3.1   Pharmacopeia Limits of Resin Obtained from *Boswellia Serrata*

Parameter	Limits
Foreign matter	NMT 5%
Total Ash	NMT 10%
Acid-soluble Ash	NMT 8%
Alcohol-soluble extractive	NLT 45%
Water-soluble extractive	NLT 28%

3.2   ETHNOMEDICINAL APPLICATIONS

B. serrata is used from the ancient time in the Indian System of drugs like a piece of writing and the Unani system. Three ancient texts; Charaka's Charaka Veda (c. B.C. 700), Susruta's Susruta Veda (c. B.C. 600), Astanga's Samgraha and Hridaya (c. 130–200 A.D.), written by Vagbhata the Elder and Vagbhata the Younger delineate anti-rheumatic activity of the plant [11]. *B. serrata* herbs utilized in the Sanskritic language name of "Gajabhakshya" because of a part of the elephant diet [12]. Ancient Ayurvedic and Unani texts narrate that the plant extracts accustomed cure varied diseases like rheumatism, skin disorder, cough, diarrhea, stomachic, diuretic, central nervous diseases, channel discharges, jaundice, etc. from the last severe centuries [7, 13]. Trendy medication and medicine power-fully focuses to its use as an anti-arthritic [14], chronic medicinal drug

[15], anti-hyperlipidemic [16], anti-atherosclerotic [17], hepatoprotective [18], and anti-carcinogenic activity [19].

3.3 PHYTOCHEMISTRY

The different species of Boswellia have about 200 phytochemicals in the oleo-gum-resin extract [6]. The gum resin is a mixture of pentacyclic triterpenoid (PTs) and sugar moieties including polysaccharides (45–60%), volatile oils (8–12%), lipophilic higher diterpenes, and triterpenes (25–35%).

Volatile oils (8–12%) predominantly constituents of Salai-guggal having monoterpenoids includes α-pinene (73.3%), cis-verbenol (1.97%), trans-pinocarveol (1.80%), borneol (1.78%), myrcene (1.71%), verbenone (1.71%), limonene (1.42%), thuja-2,4(10)-diene (1.18%), phellandrene, cadinene, and p-cymene and a small amount of diterpenes (Figure 3.4). α-pinene (73.3%) is the major chemical constituent of monoterpenoids [20, 21].

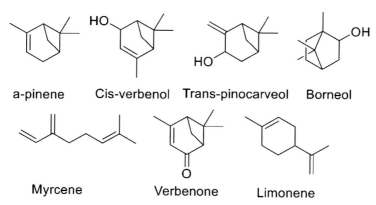

FIGURE 3.4   Structures of volatile components present in the oleo-gum-resin extract of *Boswellia serrata*.

Higher terpenoids (25–35%) comprising mainly boswellic acids (BAs) derivatives, ursane type of pentacyclic triterpenes possess potent pharmacological action. The six major BAs are α- and β-boswellic acids (BA, 10–21%), acetylated α and β-boswellic acids (ABA, 0.05–6%),

11-keto-β-boswellic acid (KBA, 2.5–7.5%), and 3-O-acetyl-11-keto-β-boswellic acid (AKBA, 0.1–3%) are present in all Boswellia species but in varying quantities [4, 22]. Boswellic acid is a chemical constituent of *B. serrata* among them AKBA has shown potent activity in the treatment of inflammation [23].

(1)

Compound Name	R_1	R_2
β-boswellic acid (BA)	α-OH	H
Acetyl β-boswellic acids (ABA)	α-OAc	H
11-keto-β-boswellic acid (KBA)	α-OH	=O
3-O-acetyl-11-keto-β-boswellic acid (AKBA)	α-OAc	=O

(2)

Compound Name	R_1	R_2
α-boswellic acids (BA)	α-OH	H
Acetyl α-boswellic acids (ABA)	α-OAc	H

Higher terpenoids BAs first isolation was reported in 1898 by Tschirch et al. [24]. Besides, boswellic acid it found that plant contains 3-O-acetyl-9,11-dehydro-β-boswellic acid (3), α-amyrin (4), 3-hydroxy-urs-9,11-dien-24-oic acid (5), 3α-hydroxy-tirucall-8,24-dien-21-oic acid (6), 3α-acetoxy-tirucall-8,24-dien-21-oic acid (7), 3β-hydroxy-tirucall-8,24-dien-21-oic acid (8), and 3-keto-tirucall-8,24-dien-21-oic acid (9) [25]. The isolation of a new diterpene alcohol, serratol (10) is responsible for the typical odor of the gum. Further presence of lupane (11), 3α-hydroxy-lup-20(29)-en-24-oic acid (12) [26], 3α-acetoxy-lup-20(29)-ene-24-oic acid (13) [27], and β-sitosterol (14) also reported from *B. serrata* (Figure 3.5). Polysaccharides (45–60%) or carbohydrates present in *B. serrata* are the acid-hydrolyzed products including arabinose, xylose, and galactose. It is reported that the presence of sugars such as arabinose and galactose, together with hexuronic acid and a polysaccharide, 4-O-methyl-glucrono-arabinoglactam from *B. serrata* [28].

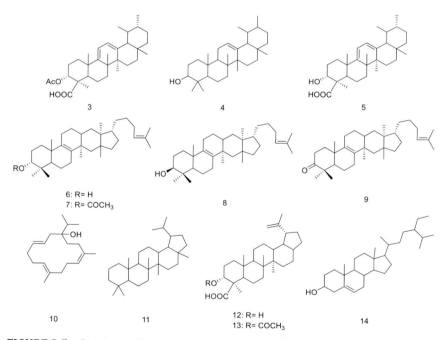

FIGURE 3.5   Structures of terpenoids present in *Boswellia serrata*.

3.4 ISOLATION OF BOSWELLIC ACIDS (BAS)

Oleoresin gum extract (BSE) composed of various boswellic acid derivatives. The higher class of triterpenoids is responsible for the therapeutic
activity and desired pharmacological response. The isolation process of
bioactive molecule BA and KBA from the alcoholic extract of gum resin
which produces a higher pure compound [4]. A German scientist Jauch
gave a simple method to synthesize BAs and their derivatives [29]. The
process of extraction was improved further using hydroalcoholic solvents,
specifically ethanol, to produce a standardized BA fraction comprising
70–90% of total Bas (Figure 3.6). The later process is the cost-effective
and eco-friendly process for purification of the BA extract [30].

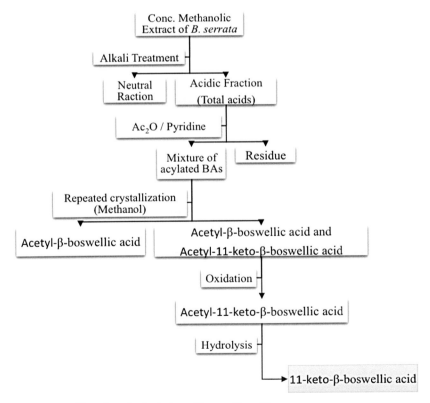

FIGURE 3.6 Flowchart for isolation of boswellic acids.

3.5   PHARMACOLOGICAL ACTIVITY

The extract (BSE) shows pharmacological activity by targeting various transcription factors (NKκB, PPAR-γ), kinases (ERK1/2, p 38 MAPK, IKKa/b, CDK-2, CDK-4), enzymes (5-LO, caspase, topoisomerase-I, and topoisomerase-II, HLE), growth factors (VEGF, PDGF, and TF) and receptors (Figure 3.7).

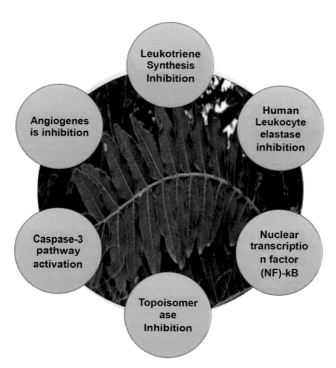

FIGURE 3.7   Proposed mechanism of action of *Boswellia serrata.*

3.5.1   ANTI-INFLAMMATORY ACTIVITY

BAs from *B. serrata* natural resin extract (BSE) exerts potent anti-inflammatory activities that are delineated in ancient texts. BAs are the precise non-redox inhibitors of leukotriene synthesis by interacting pro-inflammatory accelerator 5-lipoxygenase (5-LO) together with 5-hydroxyeicosatetraenoic acid (5-HETE) and leukotriene B_4 (LTB_4). It was rumored that BAs inhibit leukotriene synthesis by interacting 5-LO, however, not exert any result via 12-lipoxygenase or enzyme and unable to forestall peroxidation of arachidonic acid by iron or ascorbate [31] (Figure 3.8).

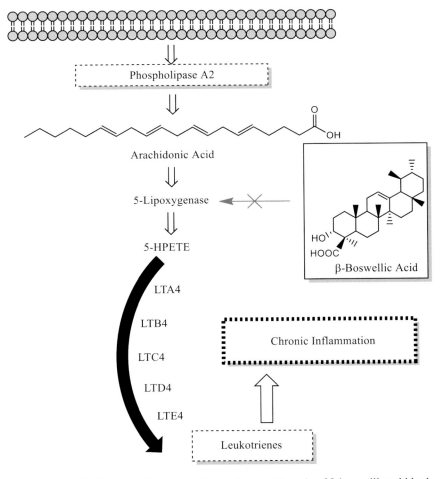

FIGURE 3.8 Mechanism of chronic inflammation and the role of β-boswellic acid in the inhibition of 5-LOX.

The methanolic extract of *B. serrata* possesses anti-inflammatory drug activity by inhibiting the cytokines like tumor necrosis factor-alpha (TNF-α), interleukin-1β (IL-1β) via nuclear transcription issue (NF)-κB, that answerable for regulation of cytokines gas and agent activated super-molecule enzyme (MAPK) in human peripheral blood mononucleate cells and mouse macrophages [7] (Figure 3.9). A study reported that animal disease strangled 25–46% paw hydrops in rats proved anti-inflammatory drug property. BAs caused a vital reduction of inflammatory symptoms additionally in papaya latex-induced rat paw inflammation (35% inhibition) [32, 33] or in 12-O-tetradecanoylphorbol-13-acetate-induced ear inflammation in mice.

Another study showed that extracts of Salai-guggal strangled the infiltration of polymorphonuclear leukocytes (PMNL) in gum iatrogenic rat hind paw hydrops model [34]. Recently, it was determined that madly animal disease additionally exerts similar effectiveness of mesalazine within the treatment of Crohn's disease [35]. It is additionally reported that by inhibiting the human corpuscle enzyme (HLE), pentacyclic triterpenes play a role in the treatment of assorted hypersensitivity base and inflammatory disorders like pancreatic fibrosis, acute metastasis distress syndrome, bronchitis, respiratory disorder, nephritis, and rheumatic inflammatory disease [20].

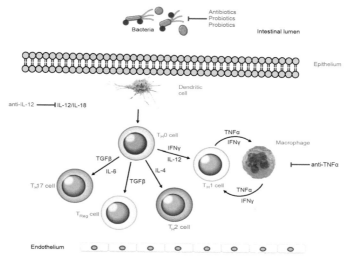

FIGURE 3.9 Proposed pathogenesis of Crohn's disease with the target site of boswellic acid derivatives for pharmacological invention.

3.5.2 ANTICANCER ACTIVITY

The extract obtained from *B. serrata* primarily containing 3-O-acetyl-β-boswellic acid, KBA, and AKBA possess *in-vitro* anticancer activity by inhibiting deoxyribonucleic acid, ribonucleic acid, and supermolecule synthesis in human cancer of the blood HL-60 cells. The distinguished activity of AKBA changes morphological characteristics of HL-60 indicates cells underwent caspase-mediated cell death [36, 37]. The BAs (AKBA, KBA) exerts their cytotoxic effects by inhibiting topoisomerase-I and topoisomerase-II, thereby resulting in complete inhibition of cell growth and proliferation established by inducement caspase-mediated cell death via a caspase-8 dependent pathway in human cancer of the blood, colon, hepatocellular carcinoma, and in varied different neoplastic cell lines like NB4, SKNO-1, K562, U937, ML-1, and HL-60 cells [38, 39]. It was evaluated for the prime role of AKBA on the expansion of glandular cancer. The analysis expressed that AKBA inhibits ontogeny induced by activation of VFGFR2 and mTOR signal pathways [40].

BAs act on cancer cells (CRC) by upregulation of the expression of two acknowledged neoplasm restrictive miRNAs like let-7 and miR-200 families (CDK6, vimentin, and E-cadherin) [41]. Oral administration of AKBA decreases the neoplasm volume of human CRC tumors beside inhibits the enlargement and metastasis of human CRC tumors. These acts through the suppression and downregulation of varied cancer-associated biomarkers like nuclear factor-κB (NF-κB), growth survival like bcl-2, bcl-XL, a matter of caspase-mediated cell death (IAP-1) and surviving; proliferative (Cyclin D1), invasive (intercellular adhesion molecule one and matrix metalloproteinase-9); invasive and angiogenic C-X-C receptor four and tube-shaped structure epithelium protein. The apoptotic result of AKBA was confirmed by the activation of caspase-3 and cleavage of poly-(ADP ribose) enzyme (PARP). It is conjointly rumored that AKBA downregulated CXCR4 expression carcinoma cell [7] (Figure 3.10).

3.5.3 ANTI-ARTHRITIC ACTIVITY

From the age, Ayurvedic drugs system extract of *B. serrata* accustomed to treat rheumatoid diseases. Within the anti-arthritic studies, it has shown 34–49% inhibition of paw swelling on eubacteria adjuvant-induced polyarthritis in rats [8]. It additionally showed that a mixture of boswellic acid exhibited 45–67% anti-arthritic activity [42]. By inhibiting the 5-LOX enzyme, the associated inflammatory marker in degenerative arthritis gets decreased. The salai-guggal extract helps to preserve the structural components of the joint gristle. 5-Loxin, a BAs extract enriched with AKBA exerts potential medicine action by inhibiting pro-inflammatory protein mediators and inhibits HLE plays a necessary role in the treatment of the autoimmune disorder.

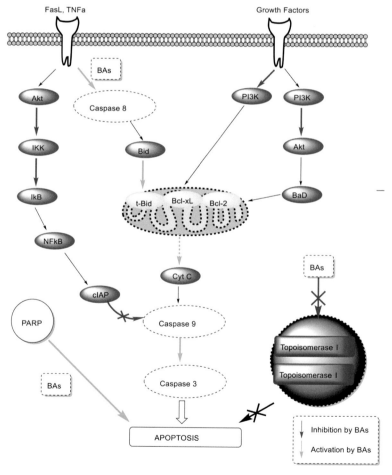

FIGURE 3.10   Schematic representation of the apoptotic signaling pathway by boswellic acid.

3.5.4 ANTI-ULCER ACTIVITY

Research showed that crude ether and binary compound extract of baccalaureate considerably show anti-ulcer activity at the dose of 250 mg/kg marked by the reduction ulceration index in peptic ulceration [43]. It was reported that animal disease showed higher activity than sulfasalazine in lesion patients at the dose of 350 mg three times daily [44], Another study at the dose of 900 mg daily for 6 weeks showed similar results when put next to sulfasalazine. It is believed that the anti-ulcerative impact possesses by the leukotrienes (LTs) synthesis pathway [45].

3.5.5 HEPATOPROTECTIVE ACTIVITY

Alcoholic extract of Salai-guggal (AESG) causes hepatoprotection in galactosamine/endotoxin evoked liver injury in mice that exerted by a reduced level of SGOT, SGPT, aminopherase, and humor enzymes [46]. Information from the varied study reported the hepatoprotective action of Salai-guggal likely through inhibition of 5-LO activity [47]. Another study planned that the methane series extract of oleo-gum-resin of Salai possesses hepatoprotective activity in lower doses by reducing the elevated levels of humor marker enzymes and prevented the rise in liver weight [18].

3.5.6 HYPOLIPIDEMIC ACTIVITY

Lipopolysaccharide-induced NO production inhibition by water-soluble *B. serrata* plant oleo gum resin in the rat peritoneal macrophages is the proposed mechanism of hypolipidemic activity [48]. Salai gum maintains the level of the serum cholesterol and triglycerides (TG) levels of animals in optimum concentration in serum, which are fed on high cholesterol and saturated fat containing diet [49].

3.5.7 ANTI-ASTHMATIC ACTIVITY

It was reported that seventy patients suffering from the bovine spongiform encephalitis got improved from the wheezy conditions in a quandary placebo-controlled study with a potential result within the treatment of respiratory disease [50]. The clinical studies on anti-asthmatic action incontestable that inhibition of leukotriene synthesis via CysLT antagonist and 5-LO inhibition [51]. Bovine spongiform encephalitis at dose 40 mg/kg *in vitro* may be potent mast stabilizing agents that free PGD_2 and aminoalkane liable for bronchoconstriction [52]. It is additionally explicit that the bovine spongiform encephalitis inhibited catG those are liable for bronchial asthma generation and COPD.

3.5.8   PERITUMORAL BRAIN EDEMA

The European Medicines Agency (EMA) classified BSE as an associate orphan drug for the treatment of peritumoral brain hydrops at a dose of 1200 mg thrice daily. Adjuvant medical aid improves the standard of life by cut back neurologic incapacity related to peritumoral brain hydrops [53].

3.5.9   IMMUNOMODULATORY ACTIVITY

A study report that BAs create a target in antimicrobial amide LL-37, an accountable development of reaction disorders and exhibit varied action in the system [54]. BA inhibits the C2 convertase protein that has a precise role within the classical complement pathway for specific immunity and helps in the production of protein of specific categories in the immune milieu [55].

3.5.10   HYPOGLYCEMIC ACTIVITY

An Ayurvedic formulation containing *B. serrata* oleo-gum-resin has been reported to supply vital anti-diabetic activity on non-insulin dependent DM (NIDDM) in a streptozocin-elicited diabetic rat model by moving viscus gluconeogenesis, pyruvate carboxylase, and phosphoenolpyruvate carboxykinase [56].

3.5.11   DIURETIC ACTIVITY

Extract of *B. serrata* considerably shows drug activity by enhancing water and solution excretion with none signs of toxicity. Drug activity of BSE evaluated by excretion of the metal particle (Na^+), metallic element particle (K^+), chloride particle (Cl^-), and hydrogen carbonate particle (HCO_3^-) show increase in the urinary output elicited by BSE plant extract [57, 58].

3.6 FORMULATIONS

BSE preparations are used in numerous cosmetic formulation, ceremony, soaps, and creams. A branded formulation Boswellin® supported methanolic extract of the natural resin exudates of *B. serrata* factory-made by Sabinsa Corporation marketed obtainable into the United States of America and Central America in 1991 within the type of tablets and capsule indefinite quantity type. Shallaki® contains 125 mg Salai in every capsule factory-made by Chain Pharma, India, as licensed User of the Trade Mark in hand by MMI Corporation exerts potent anti-inflammatory drug activities.

A cream, Niltan®, factory-made by Dr. Reddy's Laboratories Ltd., India, consisting of a mixture of active flavor extracts (boswellin, arbutin, liquorice extract, and coriander oil during a cream base). It reduces the activity of the accelerator tyrosinase at intervals the skin, therefore decreasing the assembly of animal pigment, which ends up within the reduction of dark skin formation. Rheumatic-X® factory-made by Sunrise Herbals, India, applicable for creaky, degenerative joint disease, illness, and sciatic pain, The potent anti-inflammatory drug activity of *B. serrata* studied within the type of BSE in business preparations 5-Loxin enriched with half-hour AKBA [59]. A completely unique preparation of *B. serrata* extract, Aflapin® enriched with two-hundredth AKBA and *B. serrata* non-volatile oil showed the therapeutic result for the management of inflammatory disease [60, 61].

3.7 CONCLUSION

Traditionally it is beneficial in ailments like diarrhea, asthma, arthritis, inflammation of innards, skin diseases, ulcers, bronchitis, depression, vomiting or hemorrhage from any a part of the body. In aforesaid disorders, this herb has been pharmacologically and clinically established because of anti-arthritic, anti-diarrheal, anti-depressant, anti-asthmatic, medicinal drug, anti-convulsant properties, and helpful in inflammatory viscus diseases. Moreover, recently the opposite pharmacological activities like anti-cancer, hepatoprotective, hypolipidemic, and symptom properties also are confirmed. These activities square measures are attributable to its phytochemical constituents like boswellic acid, tannin, phenol, β-sitosterol, etc.

KEYWORDS

- *Boswellia serrata*
- boswellic acids
- interleukin-1β
- nuclear factor-κB
- pentacyclic triterpenoid
- tumor necrosis factor-alpha

REFERENCES

1. Evans, W. C., (2002). Trease and evans. *Pharmacognosy* (9th edn., pp. 553–557). Published by Saunders Elsevier.
2. Orwa, C., Mutua, A., Kindt, R., Jamnadass, R., & Simons, A., (2009). *Agroforestree Database: A Tree Reference and Selection Guide, Version 4.0.*
3. Rangari, D. V., (2004). Natural colorants and dye. *Pharmacognosy and Phytochemistry: Part II* (1st edn.). In: Career publications.
4. Shah, B. A., Qazi, G. N., & Taneja, S. C., (2009). Boswellic acids: A group of medicinally important compounds. *Natural Product Reports, 26*(1), 72–89.
5. Siddiqui, M. Z., (2011). *Boswellia serrata*, a potential anti-inflammatory agent: An overview. *Indian Journal of Pharmaceutical Sciences, 73*(3), 255–261. doi: 10.4103/0250-474X.93507.
6. Iram, F., Khan, S. A., & Husain, A., (2017). Phytochemistry and potential therapeutic actions of *Boswellic* acids: A mini-review. *Asian Pacific Journal of Tropical Biomedicine, 7*(6), 513–523.
7. Bansal, N., Mehan, S., Kalra, S., & Khanna, D., (2013). *Boswellia serrata*-frankincense (a Jesus gifted herb); an updated pharmacological profile. *Pharmacologia., 4*(6), 457–463.
8. Sultana, A., Rahman, K. U., Padmaja, A. R., & Rahman, S. U., (2013). *Boswellia serrata Roxb. a Traditional Herb with Versatile Pharmacological Activity: A Review.*
9. Sunnichan, V. G., Mohan, R. H. Y., & Shivanna, K. R., (2005). Reproductive biology of *Boswellia serrata*, the source of Salai-guggul, an important gum-resin. *Botanical Journal of the Linnean Society, 147*(1), 73–82.

10. Graves, G., (1996). *Medicinal Plants: An Illustrated Guide to More Than 180 Herbal Plants*. Bracken Books.
11. Pharmacopoeia, I., (2007). *The Controller of Publication* (p. 1). New Delhi; Ministry of health and family welfare, India.
12. Sharma, S., Thawani, V., Hingorani, L., Shrivastava, M., Bhate, V. R., & Khiyani, R., (2004). Pharmacokinetic study of 11-Keto beta-Boswellic acid. *Phytomedicine, 11*(2–3), 255–260.
13. Dhiman, A. K., & Kumar, A., (2006). *Ayurvedic Drug Plants*: Daya Books.
14. Kimmatkar, N., Thawani, V., Hingorani, L., & Khiyani, R., (2003). Efficacy and tolerability of *Boswellia serrata* extract in treatment of osteoarthritis of knee: A randomized double blind placebo controlled trial. *Phytomedicine, 10*(1), 3–7.
15. Singh, G. B., & Atal, C. K., (1986). Pharmacology of an extract of Salai-guggal ex-*Boswellia serrata*, a new non-steroidal anti-inflammatory agent. *Agents and Actions, 18*(3–4), 407–412.
16. Azadmehr, A., Ziaee, A., Ghanei, L., Huseini, H. F., Hajiaghaee, R., Tavakoli-Far, B., & Kordafshari, G., (2014). A randomized clinical trial study: Anti-oxidant, anti-hyperglycemic and anti-hyperlipidemic effects of olibanum gum in type 2 diabetic patients. *Iranian Journal of Pharmaceutical Research: IJPR, 13*(3), (1003).
17. Rajeshwari, H. P., Naveen, K., M., Babu, R. L., & Chidananda, S. S., (2015). Anti-inflammatory activity of *Vitex negundo*, *Boswellia serrata* and *Aegle marmelos* leaf extracts in LPS treated A549 cells. *International Journal of Phytopharmacy, 5*(2), 12–20.
18. Kamath, J. V., & Asad, M., (2006). Effect of hexane extract of *Boswellia serrata* oleo-gum resin on chemically induced liver damage. *Pakistan Journal of Pharmaceutical Sciences, 19*(2), 129–133.
19. Huang, M. T., Badmaev, V., Ding, Y., Liu, Y., Xie, J. G., & Ho, C. T., (2000). Anti-tumor and anti-carcinogenic activities of triterpenoid, β-boswellic acid. *Biofactors, 13*(1–4), 225–230.
20. Al-Harrasi, A., & Al-Saidi, S., (2008). Phytochemical analysis of the essential oil from botanically certified oleo gum resin of *Boswellia sacra* (Omani Luban). *Molecules, 13*(9), 2181–2189.
21. Kasali, A. A., Adio, A. M., Oyedeji, A. O., Eshilokun, A. O., & Adefenwa, M., (2002). Volatile constituents of *Boswellia serrata roxb.* (Burseraceae) bark. *Flavor and Fragrance Journal, 17*(6), 462–464.
22. Büchele, B., & Simmet, T., (2003). Analysis of 12 different pentacyclic triterpenic acids from frankincense in human plasma by high-performance liquid chromatography and photodiode array detection. *Journal of Chromatography B, 795*(2), 355–362.
23. Krieglstein, C. F., Anthoni, C., Rijcken, E. J. M., Laukötter, M., Spiegel, H. U., Boden, S. E., & Schürmann, G. J. I. J. O. C. D., (2001). Acetyl-11-keto-β-boswellic acid, a constituent of a herbal medicine from *Boswellia serrata* resin. *Attenuates Experimental Ileitis, 16*(2), 88–95. doi: 10.1007/s003840100292.
24. Tschirch, A., (1898). Examinations of the secretions. *About the Olibanum, 236*(5–8), 487–503.
25. Pardhy, R. S., & Bhattacharyya, S. C., (1978). Tetracyclic triterpene acids from the resin of *Boswellia serrata roxb*. *Indian Journal of Chemistry, 16*(3), 174–175.

26. Culioli, G., Mathe, C., Archier, P., & Vieillescazes, C., (2003). A *Lupane triterpene* from frankincense (*Boswellia* sp., Burseraceae). *Phytochemistry, 62*(4), 537–541. doi: 10.1016/s0031-9422(02)00538-1.

27. Belsner, K., Büchele, B., Werz, U., & Simmet, T., (2003). Structural analysis of 3-α-acetyl-20 (29)-lupene-24-oic acid, a novel pentacyclic triterpene isolated from the gum resin of *Boswellia serrata*, by NMR spectroscopy. *Magnetic Resonance in Chemistry, 41*(8), 629–632.

28. Malandkar, M. A., (2013). Chemical constitution of the gum from *Boswellla serrata*. *Journal of the Indian Institute of Science, 8*, 240.

29. Jauch, J., (2002). *Simple Method for the Synthesis of Boswellic Acids and Derivatives Thereof.* WO2002085921.

30. Shah, P., & Thomas, G. P., (2014). *An Improved Process for Preparation and Purification of Boswellic Acids.* In: WO2014195976A2.

31. Ammon, H. P. T., Safayhi, H., Mack, T., & Sabieraj, J., (1993). Mechanism of anti-inflammatory actions of curcumine and boswellic acids. *Journal of Ethnopharmacology, 38*(2–3), 105–112.

32. Gupta, O. P., Sharma, N., & Chand, D., (1992). A sensitive and relevant model for evaluating anti-inflammatory activity-papaya latex-induced rat paw inflammation. *Journal of Pharmacological and Toxicological Methods, 28*(1), 15–19.

33. Gupta, O. P., Sharma, N., & Chand, D., (1994). Application of papaya latex-induced rat paw inflammation: Model for evaluation of slowly acting anti-arthritic drugs. *Journal of Pharmacological and Toxicological Methods, 31*(2), 95–98.

34. Sharma, M. L., Khajuria, A., Kaul, A., Singh, S., Singh, G. B., & Atal, C. K., (1988). Effect of Salai-guggal ex-*Boswellia serrata* on cellular and humoral immune responses and leucocyte migration. *Agents and Actions, 24*(1–2), 161–164.

35. Abdel-Tawab, M., Werz, O., & Schubert-Zsilavecz, M., (2011). *Boswellia serrata. Clinical Pharmacokinetics, 50*(6), 349–369. doi: 10.2165/11586800-000000000-00000.

36. Hoernlein, R. F., Orlikowsky, T. H., Zehrer, C., Niethammer, D., Sailer, E. R., Simmet, T., & Ammon, H. P. T., (1999). Acetyl-11-keto-β-boswellic acid induces apoptosis in HL-60 and CCRF-CEM cells and inhibits topoisomerase I. *Journal of Pharmacology and Experimental Therapeutics, 288*(2), 613–619.

37. Shao, Y., Ho, C. T., Chin, C. K., Badmaev, V., Ma, W., & Huang, M. T., (1998). Inhibitory activity of Boswellic acids from *Boswellia serrata* against human leukemia HL-60 cells in culture. *Planta Medica, 64*(04), 328–331.

38. Suhail, M. M., Wu, W., Cao, A., Mondalek, F. G., Fung, K. M., Shih, P. T., & Lin, H. K., (2011). *Boswellia sacra* essential oil induces tumor cell-specific apoptosis and suppresses tumor aggressiveness in cultured human breast cancer cells. *BMC Complementary and Alternative Medicine, 11*(1), 129.

39. Xia, L., Chen, D., Han, R., Fang, Q., Waxman, S., & Jing, Y., (2005). *Boswellic* acid acetate induces apoptosis through caspase-mediated pathways in myeloid leukemia cells. *Molecular Cancer Therapeutics, 4*(3), 381–388.

40. Pang, X., Yi, Z., Zhang, X., Sung, B., Qu, W., Lian, X., & Liu, M., (2009). Acetyl-11-keto-β-boswellic acid inhibits prostate tumor growth by suppressing vascular endothelial growth factor receptor 2-mediated angiogenesis. *Cancer Research, 69*(14), 5893–5900.

41. Takahashi, M., Sung, B., Shen, Y., Hur, K., Link, A., Boland, C. R., & Goel, A., (2012). *Boswellic* acid exerts antitumor effects in colorectal cancer cells by modulating expression of the let-7 and miR-200 microRNA family. *Carcinogenesis, 33*(12), 2441–2449.

42. Singh, G. B., & Atal, C. K., (1984). Pharmacology of an extract of Salai-guggul ex-*Boswellia serrata*. *Indian J. Pharmacol, 16*, 51.

43. Zeeyauddin, K., Narsu, M. L., Abid, M., & Ibrahim, M., (2011). Evaluation of antiulcer activity of *Boswellia serrata* bark extracts using aspirin induced ulcer model in albino rats. *J. Med. Allied. Sci., 1*(1), 14–20.

44. Gupta, I., Parihar, A., Malhotra, P., Singh, G. B., Lüdtke, R., Safayhi, H., & Ammon, H. P., (1997). Effects of *Boswellia serrata* gum resin in patients with ulcerative colitis. *European Journal of Medical Research, 2*(1), 37–43.

45. Gupta, I., Parihar, A., Malhotra, P., Gupta, S., Lüdtke, R., Safayhi, H., & Ammon, H. P. T., (2001). Effects of gum resin of *Boswellia serrata* in patients with chronic colitis. *Planta Medica., 67*(05), 391–395.

46. Gerlach, U., (1965). Sorbitol dehydrogenase. In: *Methods of Enzymatic Analysis* (pp. 761–764). Elsevier.

47. Safayhi, H., Mack, T., & Ammon, H. P., (1991). Protection by boswellic acids against galactosamine/endotoxin-induced hepatitis in mice. *Biochemical Pharmacology, 41*(10), 1536–1537.

48. Pandey, R. S., Singh, B. K., & Tripathi, Y. B., (2005). *Extract of Gum Resins of Boswellia serrata L. Inhibits lipopolysaccharide Induced Nitric Oxide Production in Rat Macrophages Along with Hypolipidernic Property.*

49. Zutshi, U., Rao, P. G., Kaur, S., Singh, G. B., Surjeet, S., & Atal, C. K., (1986). Mechanism of cholesterol lowering effect of Salai-guggal ex. *Boswellia serrata* roxb. *Indian Journal of Pharmacology, 18*(3), 182.

50. Gupta, I., Gupta, V., Parihar, A., Gupta, S., Lüdtke, R., Safayhi, H., & Ammon, H. P., (1998). Effects of *Boswellia serrata* gum resin in patients with bronchial asthma: Results of a double-blind, placebo-controlled, 6-week clinical study. *European Journal of Medical Research, 3*(11), 511–514.

51. Steinhilber, D., (1999). 5-Lipoxygenase: A target for anti-inflammatory drugs revisited. *Current Medicinal Chemistry, 6*(1), 71–86.

52. Badie, B., Schartner, J. M., Hagar, A. R., Prabakaran, S., Peebles, T. R., Bartley, B., & Vorpahl, J., (2003). Microglia cyclooxygenase-2 activity in experimental gliomas: Possible role in cerebral edema formation. *Clinical Cancer Research, 9*(2), 872–877.

53. Boker, D. K., & Winking, M., (1997). Die rolle von *Boswellia*-sauren in der therapie maligner gliome. *German Medical Gazette-Medical Announcements Edition A, 94*(18), 1197–1198.

54. Henkel, A., Tausch, L., Pillong, M., Jauch, J., Karas, M., Schneider, G., & Werz, O., (2015). *Boswellic* acids target the human immune system-modulating antimicrobial peptide LL-37. *Pharmacological Research, 102*, 53–60.

55. Knaus, U., & Wagner, H., (1996). Effects of *boswellic* acid of Boswellia serrata and other triterpenic acids on the complement system. *Phytomedicine, 3*(1), 77–80.

56. Al-Awadi, F., Fatania, H., & Shamte, U., (1991). The effect of a plants mixture extract on liver gluconeogenesis in *Streptozotocin* induced diabetic rats. *Diabetes Research (Edinburgh, Scotland), 18*(4), 163–168.

57. Asif, M., Jabeen, Q., Abdul-Majid, A. M., & Atif, M., (2014). Diuretic activity of *Boswellia serrata* Roxb. oleo gum extract in albino rats. *Pak. J. Pharm. Sci., 27*(6), 1811–1817.
58. Reddy, G. K., Chandrakasan, G., & Dhar, S. C., (1989). Studies on the metabolism of glycosaminoglycans under the influence of new herbal anti-inflammatory agents. *Biochemical. Pharmacology, 38*(20), 3527–3534.
59. Sengupta, K., Alluri, K. V., Satish, A. R., Mishra, S., Golakoti, T., Sarma, K. V. S., Dey, D., & Raychaudhuri, S. P. (2008). A double blind, randomized, placebo controlled study of the efficacy and safety of 5-Loxin® for treatment of osteoarthritis of the knee. *Arthritis Research & Therapy, 10*(4), R85.
60. Berman, B., Lewith, G., Manheimer, E., Bishop, F. L., & D'Adamo, C., (2015). 48A-complementary and alternative medicine. In: Hochberg, M. C., Silman, A. J., Smolen, J. S., Weinblatt, M. E., & Weisman, M. H., (eds.), *Rheumatology* (6th edn., pp. 382–389). Philadelphia: Content Repository Only.
61. Suva, M. A., Kheni, D. B., & Sureja, V. P. (2018). Aflapin®: A novel and selective 5-lipoxygenase inhibitor for arthritis management. *Indian Journal of Pain, 32*(1), 16–23.

CHAPTER 4

Generation of Natural Pharmaceuticals Based on Microbial Transformation of Herbal Constituents

M. R. BADWAR,[1] AKSHADA A. BAKLIWAL,[2] SWATI G. TALELE,[2] and ANIL G. JADHAV[3]

[1]Department of Quality Assurance, Sandip Institute of Pharmaceutical Sciences, Mahiravani, Nashik, Maharashtra, India

[2]Department of Pharmaceutics, Sandip Institute of Pharmaceutical Sciences, Mahiravani, Nashik, Maharashtra, India

[3]Department of Pharmacognosy, Sandip Institute of Pharmaceutical Sciences, Mahiravani, Nashik, Maharashtra, India

ABSTRACT

Herbal constituents efficiently transformed by utilizing technology as biotransformation to diminish the droning process and toxicity of the synthetic compounds. This progression is facilitated by the involvement of microorganisms and their products such as bacteria, fungi, and enzymes. Microbial biotransformation or microbial biotechnology is gaining importance day-by-day and extensively utilized to generate metabolites in bulk amounts with more specificity. The present chapter elaborates on how efficiently microbial transformation use in natural products (NPs) and the results of the most recent developments in the era of natural pharmaceuticals. The scope of the chapter is to demonstrate microbial transformation as a step in the generation of natural pharmaceuticals from herbal constituents. Microbial biotransformation offers various therapeutic applications in the field of various phytoconstituents including steroid drugs and their constituents as terpenes, alkaloids, flavonoids, and polyphenols giving derivatives with improved biological activities.

4.1   INTRODUCTION

4.1.1   DEFINITION OF BIOTRANSFORMATION

Biotransformations are basic alterations in a biochemical compound by life forms/catalyst frameworks that lead to the arrangement of atoms with moderately more prominent extremity [1, 2]. This component has been created by microorganisms to adjust to ecological changes and it is valuable in a wide scope of biotechnological forms [3]. The hugest part of biotransformation is that it keeps up the first carbon skeleton subsequent to getting the items [4].

The biotransformation is of two categories:

1.   enzymatic biotransformation; and
2.   non-enzymatic biotransformation.

Enzymatic is additionally separated into microsomal and non-microsomal [5]. Enzymatic elimination is the biotransformation happening because of different catalysts present in the body. Microsomal biotransformation is brought about by compounds present inside the lipophilic layers of the smooth endoplasmic reticulum [6].

Non-microsomal biotransformation includes the chemicals which are available inside the mitochondria. Models include: Alcohol dehydrogenase in charge of digestion of ethanol into acetaldehyde and Tyrosine hydrolases chemicals, xanthine oxidase changing over hypoxanthine into xanthine and so on. Unconstrained, non-catalyzed, and non-enzymatic sorts of biotransformation are for exceptionally dynamic, flimsy mixes occurring at physiological pH. A portion of these incorporate Chlorazepate changed over into Desmethyl diazepam, Mustin HCl changed over into Ethyleneimonium, Atracurium changed over into Laudanosine and Quartenary corrosive, Hexamine changed over into formaldehyde.

Microbial biotransformation is generally utilized in the change of different poisons or an enormous assortment of compounds including hydrocarbons, pharmaceutical substances, and metals [7]. These changes can be congregated under the classifications: oxidation, decrease, hydrolysis, isomerization, and practical gatherings [8]. For quite a long time microbial biotransformation has demonstrated to be a basic device in mitigating the creation of different synthetic substances utilized in sustenance, pharmaceutical, agrochemical, and different industries.

In the field of pharmaceutical innovative work, biotransformation studies have been widely connected to research the metabolism of compounds utilizing creature models [9]. The microbial biotransformation marvel is then usually utilized in looking at metabolic pathways of medications and scaling up the metabolites of intrigue found in these creature models for further pharmacological and toxicological assessment [10].

Biocatalysis extent of study including microbial change is expanding fundamentally from restricted enthusiasm into the very dynamic zone in science today including the arrangement of pharmaceutical items. Biotransformation can be explained as the particular alteration of an unmistakable compound to an unmistakable item with basic closeness, by the utilization of natural impetuses including microorganisms like organisms. The organic catalyst can be depicted as an enzyme, or an entire, inactivated microorganism that contains an enzyme or several enzymes produced in it.

Bioconversion is another term identified with microbial change for this examination specifically. There is just a slight contrast between biotransformation and bioconversion.

A bioconversion uses the synergist action of living life forms and consequently can include a few concoction response steps. A living microorganism will be constantly delivering catalysts and subsequently, bioconversions regularly include chemicals, which are very temperamental for utilized substrates. The properties of biotransformations and bioconversions are fundamentally the same as and by and large, the terms are referred to as tradable. Then again, aging, science under zymology uses microorganisms; yeast was known to transform sugar into liquor since 1857 by the French scientist, Louis Pasteur. The biotransformation procedures have favorable circumstances conquered a portion of the inalienable issues and instances of some financially effective procedures. To use these procedures, biocatalysis research has been proposed for the country's rich normal assets for the most part with the endophytes accessible.

Biotransformation procedures are unquestionably more differing than restorative protein generation forms. There are numerous microorganism strains and proteins required to misuse the specific biotransformation potential for the bioconversion of a bunch of various substances into the ideal items, particularly new optically dynamic fundamental pharmaceutical ingredients.

The course of events compressions in the improvement cycle of pharmaceuticals, in the mix with a missing wide strain and chemical decision, bring about the way that biotransformation normally speaks to the subsequent age process decision in the assembling of a little atom pharmaceutical. Novel biocatalysts are required most importantly particularly oxidoreductases what's more, lyases for biotransformation.

Biotransformation is likewise known to agree to the green science system today. Green science is a term utilized for feasible substance modern assembling forms toward accomplishing negligible waste generation and vitality utilization. In this way, biosynthesis, and biotransformation are accepted to assume a key job in green science in the years to come [11–15].

4.1.2   ADVANTAGES OF MICROBIAL TRANSFORMATION

Numerous advantages can be acquired through microbial changes contemplates. The procedure required in microbial change may most likely be able to work at close unbiased pH, surrounding temperatures, and climatic weights. Interestingly, science frequently requires limits of these conditions which are not actually naturally benevolent and mechanically undesired. Moreover, extraordinary pH, temperature, and weight may give destructive impacts toward the workforce working the cruel techniques and may likewise influence network encompassing the territories [13]. All the more critically biocatalysts are very response explicit, enantiomer-explicit, and regio-explicit. This is chiefly and straightforwardly alluding to the synthetic structure of a compound one might need to acquire explicitly. Numerous adaptable microorganisms can be used to complete incredibly explicit transformations utilizing substrates of minimal effort.

The fundamental science responses incorporate expansion responses, end responses, substitution responses, pericyclic responses, revision responses, and redox responses. The means might be long and increasingly dull now and again as synthetic substances are effectively bothered by the sticky condition in tropical regions for example. Microorganisms have extraordinary potential for initiating numerous choices of inventive and ad-libbed chemical frameworks, which are competent for changing over new substrates. In this way, numerous investigations can be performed to a more noteworthy broaden with respect to various endophyte species toward concoction modifications of atoms and mixes of premium. The genome of a novel the cryophilic contagious animal groups can be surveyed to give quality groupings that encode for thermotolerant compounds, which are progressively steady to varieties of response temperature. Not as the name may propose physically, microorganisms are animal unbelievably little for the unaided eyes to see however convey real jobs today in the pharmaceutical industry somehow.

Microorganisms are equipped for creating interesting catalysts which are steady toward warmth, antacid, and acids. One of the investigations done was with respect to hyperthermophilic archaeon *Pyrobaculum calidifontis* VA1 which delivered a thermostable esterase [16, 17].

Their little size has by a wide margin the biggest surface-to-volume proportion in correlation with some living beings. Subsequently, this enables them to expand their metabolic rates due to a high trade of particles and metabolites through their surface. With the correct development conditions, microorganisms develop exponentially.

Microorganisms are competent to create an extraordinary assortment of compounds in a brief timeframe because of its regular trademark to duplicate. It is additionally conceivable to get and develop microorganisms that can get by under extraordinary situations, for example, low or high temperatures as well as acidic or antacid conditions. Microbial change can make doable responses that are not liable to be completed by customary engineered techniques. Likewise, endophytes may deliver normal, biodegradable mixes.

4.1.3   PROCEDURE FOR BIOTRANSFORMATION

Vegetative cells, spores, resting cells, compounds, and immobilized cells/ proteins are commonly utilized for microbial change. In procedure with developing societies, the strain utilized is developed in an appropriate medium and a concentrated substrate arrangement is included after reasonable development of the way of life (6–24 h). A variation of this technique is to utilize an extremely enormous inoculum and to include the concentrated substrate promptly without taking into account a development period. Emulsifier, for example, Tween or solvents (water-miscible and low dangerous (ethanol, $CH_3)2CO$, dimethylformamide, dimethyl sulfoxide) might be utilized to help solubilize ineffectively dissolvable mixes. In some steroid changes, the substrate is included and changed over in fine crystalline structure. These alleged pseudo-crystalline maturations can be done with generally high groupings of the substrate.

For the biotransformation of lipophilic materials, it is conceivable to utilize a polyphase framework. The watery stage containing the cell material or the chemical is overlayed with the water-immiscible liquid stage where the substrate has been broken up. The substrate passes gradually into the watery stage and as the change response continues, the item goes once again into the dissolvable stage. Now and again, the real change happens at the interface of the fluid and dissolvable stages.

Change responses in huge scale gear are completed under sterile conditions in circulated air through and mixed fermenters, the transformation procedure being checked chromatographically or spectroscopically. The procedure is ended when a most extreme titer is come to. The upkeep of sterile conditions is fundamental since defilement can prompt the development of defective transformation items and can likewise take part in the all-out breakdown.

Resting cells can be utilized on the off chance that if chemical enlistment by the additional substrate isn't fundamental. This has an impressive favorable position that development restraint by the substrate is dispensed with. High cell densities, which advance expanded profitability, might be utilized; simultaneously danger of defilement is diminished. Since the change response happens transcendently in the cushion arrangement, the recuperation of the item is generally simple. Various change procedures utilize immobilized cells, offering a favorable position that the procedure can be done ceaselessly and the phones can be utilized again and again. Immobilized bacterial cells, which catalyze one-arrange or multi-organize responses, are by and by utilized monetarily in the generation of aspartic corrosive, L-alanine, and malic corrosive.

The final results of change responses are found extracellular and may happen in either broke up or suspended structure. The cell material is then washed tediously with water or natural solvents so as to segregate the response item which can be adsorbed to the cells. Contingent upon the dissolvability of the item, recuperation is performed by adsorption to particle exchangers, by precipitation as the calcium salt, by extraction with suitable solvents or for unstable substances, by direct refining from the medium [18–21].

White biotechnology includes the utilization of microbial biotransformation for creating results of intrigue. Living cells, for example, microbes, filamentous growths, creatures, plants, green growth, yeast, and actinomycetes are utilized.

The microbial cells are perfect choice for biotransformation because of specific reasons:

1. **Surface-Volume Proportion:** Microbial biotransformation has a high surface-volume proportion.
2. **Development Rate:** Higher development rate of microbial cells diminishes the time of biomass change.
3. **Digestion Rate:** Higher rate of digestion in microorganisms prompts a productive change of substrate.
4. **Sterility:** It is simpler to keep up sterile conditions when microorganisms are utilized [22].

Microbial biotransformations are an assortment of biochemical responses to change the structures of the phytochemicals and natural combinations, by abusing microorganisms and their disconnected proteins, to build up an assortment of helpful constituents, through regio-stereoselectivity responses.

Biotransformation can be arranged into two unique approaches, the one that includes the change of substrates that are totally unusual to the specific framework, which called "Xenobiotics," and the bio artificially coordinated biotransformation in which the substrate endures a formal connection to a characteristic biosynthetic intermediate. In Biotransformation forms, a perfect entire cell microorganism or isolated enzyme frameworks can be utilized, and each methodology has its preferences and burdens. For example, clean compounds in bio-catalysis could have selectivity for specific criticisms, basic framework and procedures, and better acknowledgment of cosolvents used to disintegrate low-water dissolvable substrates. Then again, catalyst detachment and purifying are genuinely exorbitant and requirements time, and ordinarily, it is harder to perform responses that need more than one compound. Human has regularly utilized microbial bio-catalysis since a great many years prior to the bread-making, dairy items, and mixed beverages. Literature survey revealed that

Scientistput the main logical bases for the microbial change applications, when he utilized an unadulterated culture of Bacterium; xylinium is utilized to change the liquor to acidic corrosive. Thusly, a few microbial changes' examinations have been completed, which demonstrated that a one-advance method may deliver an exceptional item. Microbial biotransformation by the entire cell microorganisms is frequently invaluable when contrasted with separated proteins; it is regarded financially and biologically an aggressive apparatus for the biotechnological experts looking for new strategies to fabricate clean profitable synthetic concoctions, pharmaceutical, and agrochemical mixes. Microbial change has been broadly utilized, to make new and helpful metabolites of practically all classes of terpenes, Steroids, and natural concentrates, for example, tea separates as a substitute of substance blend for the readiness of pharmacologically dynamic mixes. Biotransformation can be a few times the main unsurprising method to yield explicit mixes, for example, the hydroxylation of the non-enacted carbon particles. In the writing, there are various models and instances of biotransformation coordinating to the creation and partition of chiral natural compounds.

The entire cell microorganism's profile catalysis gives a cumbersome accumulation of enzymatic activity and selectivity, for example, the oxygenases, which are difficult to separate, and need broad and exorbitant cofactor prerequisites. Also, utilizing entire cells may give different catalysts, e.g., oxidoreductases or hydrolytic compounds that can have substrate specificities and item selectivity. All things considered, the finding of the fitting microorganism to play out the most loved biotransformation response is as yet a major test. Consequently, customary screening of a progression of microbial strains is as yet the most usable practice.

Usually, the significance of biotransformation advancements by and large and microbial changes specifically can basically demonstrate the accompanying three purposes:

1. Restorative herb maturation preparing either by microorganisms or plant cells, so as to collect a mass of optional metabolites of the focused on the plant;
2. Building a microbial model for metabolic components of the homegrown drug. Toward understanding the suitable therapeutic metabolites in human, this since microorganisms-chemicals could

separate Xenobiotics comparably as it occurs by mammalian compounds, e.g., hydroxylation, acetylation, N-dealkylation; and

3. Assembling and modifying the compelling constituents of the phytomedicine, augment the regular items accumulation, and to draw the courses of medications combination.

So far, huge work has been sophisticated in biotransformation and various broad audits have been distributed. As a compliment, this article will audit the latest discoveries which just cover the changes of some normal items separates and terpenoids that are increasingly identified with the current research.

4.2 BIOTRANSFORMATION OF NATURAL PRODUCTS (NPS) AND EXTRACTS

The hydrolytic and reductive limits of microorganisms (microscopic organisms, yeast, and parasites) have been distinguished decades prior, and right now, they are utilized in the starter and assembling responses. Diverse bioactive phytochemicals and natural items have been presented to the microbial bio-catalysis as an endeavor to discover further enthusiastic and less lethal products.

4.2.1 HONGQUEXTRICATE

It is a reformed item that is fermented on rice by Monascuspurpureus W., and it has been devoured as nourishment and prescription in China for hundreds of years back. This item has a few biodynamic segments with lower hyperglycemia, lower hyper cholesterol, lower hyperpiesia, antimicrobial, and hostile to tumor exercises, and various related bio-dynamic mixes have been separated from Hongqu, for example, monascorubin, monacolins, notalin, and ergosterin [23].

4.2.2 GREEN TEA

Green tea (*Camellia sinensis*) and yerba mate separates have been naturally changed. These two plants have some bounty constituents of the polyphenolic aggravate that expected to contribute in the medical advantages of tea. Biotransformation of their concentrates with tannase has improved the cell reinforcement capacity to 55% and 43% for the green tea and the yerba mate correspondingly [24].

4.2.3   BIOTRANSFORMATION OF TRITERPENES

Triterpenes naturally expected to be a multipurpose accumulation of terpenes. Oleanane, ursane, lupane, and dammarane-euphane carbon skeletons expected to be the best imperative triterpenoids structures. Terpenoids have been accounted for to have various natural impacts like calming, hepato defensive, pain-relieving, antimicrobial, and hostile to mycotic, virostatic, invulnerable modulatory, and tonic impacts. Then again, some triterpenoids show a few inconveniences, for example, the hemolytic and cytostatic attributes that utmost their pharmacological practice, in expansion to that some of them are ineffectively water solvent, which fundamentally constrains its application. As outcome biotransformation of these triterpenes turned into a promising noteworthy system to beat a portion of the triterpenes limitations and to grow the scope of usable [25].

4.2.4   BETULIN

Feng et al. in the latest article have researched the biotransformation of betulin to betulinic corrosive by Cunninghamellablakesleeana cells and the LC-MS examination demonstrated that betulin could be changed over into at any rate five subsidiaries from refined C. blakesleeana cells, betulinic corrosive appeared to be the most noteworthy.

4.2.5   ASIATIC CORROSIVE AND ASIATICOSIDE

As of late, an examination gathering has distributed their biotransformation trials of Asiatic corrosive; Asiatic corrosive considered one of the major significant triterpenes of Centellaasiatica that shows wound mending, hostile to aggravation and enemies of tumors movement. Different kinds of microorganisms have been utilized in these trials which managed numerous subsidiaries.

Asiaticoside, additionally one of the major triterpenes saponins in Centellaasiatica, and was economically utilized as an injury recuperating specialist, owing fundamentally to its intense calming impacts some asiaticoside subsidiaries have solid neuroprotective impacts against beta-amyloid-actuated neurotoxicity by hostile to apoptotic and against oxidative damage components. Oxy-asiaticoside a transitional subordinate of asiaticoside and was utilized in for the tuberculosis treatment and wound injuries. Asiaticoside was changed by the removed compounds delivered by *Fusarium oxysporum* to derhamno-degluco-asiaticoside and derhamno-asiaticoside. In any case, as far as we could possibly know, up to this date there are no reports on the microbial change of Asiaticoside [26].

4.2.6   MICROBIAL TRANSFORMATION OF TERPENES

Terpenes have consistently been in the focal point of examinations because of their wide applications in the Flavor and aroma industry, just as on account of their potential for further biotechnological advancements as pharmaceutical specialists and bug sprays. Terpenes have an assortment of jobs in interceding adversarial and gainful cooperations among living beings.

There is a great deal of information in the writing portraying the capacity of explicit microorganisms to perform microbial changes of terpenes like Mucorsp. *Aspergillus niger* societies, and so forth.

There are surveys on microbial change of monoterpenes and triterpenes. The microbial changes of terpenoids connected in people drug and of enthusiasm for drug stores are likewise explored in the writing like antimalarial ones, ent-kaurane diterpenes, Schisandraceae, and oleanane triterpenoids, taxanes, and so forth. Here we will talk about microbial changes of terpenes, which manage the cost of mixes with improved natural movement.

Taxol is a normally happening diterpenoid broadly connected as an amazing enemy of malignant growth tranquilizes. In excess of 500 microorganisms were screened for their capacity to accomplish valuable biotransformation of taxol/cephalomannine and *Streptomyces* sp. Mama 7065 was chosen because of its capacity for the development of hydroxy-subordinates with essentially upgraded attributes against human tumor cell lines than the particular substrates [28–31].

4.2.7 *MICROBIAL TRANSFORMATION OF ALKALOIDS*

Alkaloids speak to a different gathering of plant characteristic items with variable substance structure. They are utilized for a considerable length of time as a result of the wide assortment of their physiological impact.

The communications of microorganisms with alkaloids are of extraordinary intrigue. The information on microbial changes of alkaloids gathered up to 2000–2001 were looked into by Abraham and Spassov while simultaneously Rathborne and Bruce underlined in their survey paper on the designing biocatalytic courses for the creation of semisynthetic sedative medications. Morphine and codeine were changed into intense pain relieving hydromorphone and the gentle pain-relieving/antitussive hydrocodone, individually, by recombinant *E. coli.*

Demethylations, oxidations, and decreases of morphine alkaloids were performed with distinctive parasitic strains, *Cunninghamella echinulata* being the best one.

Rhizobium radiobacter was accounted for to hydroxylate codeine to its C-14 subordinate. This change response is of significance for the creation of medications showing pain-relieving, antitussive, and opiate rival attributes like oxycodone. Veratrum alkaloids are a gathering of strong hypotensive operators that lower circulatory strain by reflex concealment of the cardiovascular framework. Lü et al. revealed biotransformation of vermitaline (verazine type steroidal alkaloid segregated from the underlying foundations of *Veratrumda huricum* and one of the most widely contemplated) by *Cunninghamella echinulata* into four metabolites, three of which being new mixes.

The steroidal alkaloid dictyophlebine (strong cholinesterase inhibitor) from the plant *Sarcococca hookeriana* BAILL was changed by *Rhizopus stolonifer* into three polar subordinates, one of which uncovered higher inhibitory movement than that of the parent compound.

The antimalarial property of cinchona bark and the resulting segregation of its dynamic compound, quinine, have assumed a crucial therapeutic job in human culture for more than 300 years.

The brooding of cinchona alkaloids with the endophytic *Xylaria* sp. segregated from *Cinchona pubescens* (Rubiaceae) prompted the arrangement of three subsidiaries, quinine 1-N-oxide, quinidine 1-N oxide, and cinchonine 1-N-oxide, which uncovered pitifully hindering impact on the expansion of the intestinal sickness pathogen *Plasmodium falciparum*, a chloroquine-safe strain. Ruscogenin is a steroidal glycoside removed from ruscus roots. In Europe, the roots and stems of the Ruscusaculeatus have been utilized for quite a long time. Later clinical perceptions uncover the vasculoprotective and phlebotonic properties of butcher's sweeper based arrangements. Ruscogenin is a low water solvent, which limits its application; however, a subordinate with higher water dissolvability has been acquired through microbial change [32, 33].

4.2.8 *MICROBIAL TRANSFORMATIONS OF FLAVONOIDS*

Flavonoids are plant metabolites with organic capacities extending from a tinge of Blossoms as a visual sign that pulls in pollinators and security from bright radiation and phytopathogens to interest in pressure reactions. As indicated by Ren et al., Flavonoids are a gathering of more than 4000 polyphenolic mixes, which have a typical phenylbenzopyrone structure (C6-C3-C6) and are classified by the immersion level and opening of the focal pyranring.

The microbial change systems for the creation of flavonoids have pulled in impressive intrigue since they permit yielding of novel flavonoids, which don't exist in nature.

The accomplishment of microbial glycosylation prompted huge development in biotechnological glycosylation of flavonoids.

The principle capacity of glycosylation procedures are adjustment, detoxification, and solubilization of substrates [34, 35].

4.3 VARIOUS APPLICATIONS OF MICROBIAL BIOTRANSFORMATION

4.3.1 *TRANSFORMATION OF STEROIDS AND STEROLS*

Steroids comprise a characteristic item class of intensifies that is generally dispersed all through nature present in bile salts, adrenal-cortical, and sex-hormones, creepy-crawly shedding hormones, sapogenins, alkaloids, and a few anti-toxins. In 1937, the primary microbial biotransformation of steroids was done. Testosterone was created from dehydroepiandrosterone by utilizing *Corynebacterium* sp. In this manner, cholesterol was delivered from 4-dehydroeticholanic and 7-hydroxycholestrol utilizing *Nocardia* spp.

All steroids have a similar essential structure, a cyclo-pentane perhydrophenanthrene which comprises of four intertwined rings. Cortisone is helpful on account of its calming activity against rheumatoid joint pain and skin ailments .Cortisone converted to prednisolone by dehydrogenation or microbial transformation of cortisol to prednisolone may occur by *Corynebacterium simplex* which have the property of notably expanded calming impact. The steroid particle has a few hilter kilter focuses and it makes the absolute blend of steroid troublesome. Starter inquire about on the 11 alpha-hydroxylation of progesterone indicated the likelihood of the microbial presentation of oxygen into the steroid core in a site explicit and stereospecific way without earlier enactment. These responses functioned admirably and financially savvy generation of cortisone wound up conceivable. The above microbial advance responses are of extraordinary monetary essentialness. Progesterone change of a C-19 steroid is utilized modernly in the creation of testosterone and estrogen and the microbial lack of hydration of ring and is utilized in estrogen generation.

The breakdown item 3-hydroxy-9,10-secoand rostatriene-9,17-dione is delivered from cholesterol through an opening of the B ring, with the creation of two valuable moderate items, androstenedione, and, androstadiene dione with the assistance of a compound named as Arthrobacter simplex [36, 37].

4.3.2 TRANSFORMATION OF NON-STEROID COMPOUNDS

4.3.2.1 DIHYDROXYACETONE FROM GLYCEROL

Dihydroxyacetone has bunches of criticalness. It is utilized in creams and beauty care products. Different acidic corrosive microscopic organisms, for example, *Gluconobacter melanogenus* utilized in this microbial change.

4.3.2.2   PROSTAGLANDINS (PGS)

These are unsaturated greasy C-20 unsaturated fats hormones. These contain different therapeutic importance, for example, PGE-2 utilized as preventative, PEG-2 utilized for the easing of the agony of labor, PEG-1 for the treatment of inborn heart disappointment, and furthermore for the treatment of stomach related infections. These prostaglandins (PGs) can be delivered from unsaturated fats by microbial change with pathogenic growths, for example, Cryptococcus neoformans.

4.3.2.3   L-ASCORBIC CORROSIVE (NUTRIENT C)

The procedure for the generation of L-ascorbic corrosive is called the Reichstein-Grussner blend. This procedure of microbial change comprises of a few stages and this L-ascorbic corrosive is utilized in nutrient planning or as a cancer prevention agent in sustenance make. The oxidation arrange from D-sorbitol to L-sorbose is completed by *Acetobacter suboxydans* in a submerged procedure at 30–35C with fiery blending and air circulation. This procedure is completed constantly in two phases; the initial step includes the oxidation of glucose by Erwinia species to 2,5-diketo-D-gluconic corrosive by means of D-gluconic corrosive and 2-keto-D-gluconic corrosive. The subsequent advance includes a decrease of 2,5-DKG to 2-keto-1-gluonic corrosive which is catalyzed by a Corynebacterium animal variety.

4.3.3   TRANSFORMATION OF ANTIBIOTICS

The microbial change of existing anti-infection agents has been finished with the goal of growing new, adjusted, and improved antimicrobials which contain numerous characteristics like decreased harmfulness, expansive antimicrobial range, upgraded oral adsorption, less safe/unfavorably susceptible impacts. By and large, any change step causes a halfway or complete inactivation of anti-microbial. A few runs of the mill instances of the numerous potential responses are given here:

1. **Roundabout Change:** In controlled biosynthesis, adjusted anti-infection agents are delivered within the sight of inhibitors or altered

forerunners in the medium. For instance, *Streptomyces parvulus* produces two new actinomycins in which proline is supplanted by cis-4-methylprolin. New mixes have been discovered when freaks obstructed in the blend of a specific anti-infection were utilized. Mutational amalgamation can prompt the age of just a few improved antimicrobials, for example, 5-epi-sisomicin which has demonstrated its adequate viability to experience clinical preliminaries.

2. **Direct Change:** Hydrolysis of the useful gatherings prompted inactivation of the anti-toxins. However, on account of lankacidin-C-14-butyrate, a bioconversion item framed from lankacidin C and methyl-butyrate by Bacillus megaterium IFO 12108, improved antimicrobial action with lower lethality was acquired [38].

4.3.4   TRANSFORMATION OF PESTICIDES

Specialists for plant malady and bug control are essential for the survival of the total populace. The high solidness of the mixes utilized is imperative for the vector control programs; however, this steadiness negatively affects the earth.

In this viewpoint, microbial change is of intrigue not for the generation of new dynamic specialists, however for the best conceivable detoxification of the earth. This includes enzymatic changes of xenobiotics. Expulsion of xenobiotics from the biological system can be practiced through different instruments:

1. **Metabolism:** Xenobiotics can fill in as substrates for microbial development and vitality creation. A complete breakdown of certain substrates to carbon dioxide and water. For instance, herbicide dalapon (a chlorinated unsaturated fat) is changed over by Arthrobacter species into pyruvate.

2. **Cometabolism:** This typically causes a negligible alteration of particles, which may result in either a decline or an expansion in lethality. Microorganisms engaged with it don't acquire vitality from the change response and require another substrate for the development. A complete breakdown of a compound can be accomplished through the consolidated activity of various creatures. For

instance: Dehalogenation responses are significant cometabolism responses which may make pesticide atoms open for further break-down. A few mixes, for example, chlordecone, a hexachlorocy-clopentadiene subordinate with astounding bug spray impacts, are not effectively assaulted by microorganisms on account of their confounded structure and high level of halogenation.

The process of biotransformation utilized much more for reduction of contaminants and process called as bioremediation. The catabolic decent varieties of organisms are been used for the bioremediation of an enormous scope of mixes including polyaromatic hydrocarbons (PAHs), pharmaceutical substances, radionuclides, hydrocarbons (for example, oil) and polychlorinated biphenyls (PCBs). The principle-exasperating pattern is the danger risk these poisons posture to the general wellbeing. A significant development in the atomic systems, bioinformatics, genomics, and metagenomics are being utilized to enhance the change of xenobiotic mixes. Albeit most life forms have detoxifying capacities, organisms particularly microorganisms has a significant part to play in this proce-dure. High development rate, metabolic decent variety, and level quality exchange help them to develop and modify as indicated by the consistently changing natural conditions.

Both oxygen-consuming and anaerobic bacterial genera found have been related with the biotransformation of a wide scope of xenobiotic synthetics. *Bacillus, Pseudomonas, Escherichia, Rhodococcus, Gordonia, Moraxella*, and *Micrococcus* are individuals from the high-impact genera while the anaerobic kinds incorporate *Methanospirillum, Pelatomaculum, Syntrophobacter, Desulfotomaculum, Syntrophus, Desulfovibrio*, and *Methanosaeta. Mycobacterium vaccae* have been showing the capacities to catabolize $(CH_3)_2CO$, cyclohexane, styrene, benzene, ethylbenzene, propylbenzene, dioxane, and 1,2-dichloroethylene. Pseudomonas and Bacillus are known to debase PCB in all respects productively. A few strains of Pseudomonas, Acetobacter, and Klebsiella have additionally had the option to bio-fix cancer-causing azo mixes. It is discovered that Pseudomonas BCb12/1 and BCb12/3 have remarkable corruption capaci-ties of low ethoxylated NPnEO (non-phenol Polyethoxylates).

Anaerobic methanogens are essentially engaged with the corruption of the Phthalate compound. As of late *Cunninghamella elegans, Pseudo-monas knackmussii* and *P. pseudoalcaligenes* KF707 have demonstrated the capacity to biotransform potential poisons bearing the pentafluorosulfanyl (SF5-) useful gathering [39].

4.3.5   PETROLEUM BIOTRANSFORMATION

Petroleum hydrocarbons are the main impetus of our industry and day-by-day life. Nonetheless, one of the significant concerns nowadays is the hydrocarbon pollution coming about because of the exercises identified with the petrochemical business. Breaks and coincidental discharge are the fundamental wellsprings of soil and water contamination. Oil is cancer-causing and neurotoxic for every single living structure. So different mechanical and synthetic strategies have been utilized for soil and water remediation. Yet, bioremediation utilizing organisms is the most proficient technique to detoxify the poisons since it prompts total mineralization and is practical.

Numerous amphibians and marine microflora have been appeared to assume a significant job in the biodegradation of oil slicks by separating oil contaminants into nontoxic structures. Biotransformation of oil hydrocarbons is done principally by microscopic organisms, yeast, and parasites. Jones et al. have detailed the nearness of biodegraded oil-determined fragrant hydrocarbons in marine dregs. Daugulis and McCracken have revealed the debasement of PAHs by Sphingomonas [40].

4.4   EXAMPLES OF MICROBIAL TRANSFORMATION OF SOME BIOACTIVE PHYTOCHEMICALS

1. **Phytochemicals:** Artemisinin (triterpenoid).
   - ➤ **Origin and Bioactivity:** *Artemisia annua* Antimalaria.
   - ➤ **Microbial Catalyst:** Mucor polymorphous, *Aspergillus niger*, Cunninghamellaechinulata, and others.
   - ➤ **Transformed Products:** 3β-hydroxyArtemisinin and 3β-hydroxydeoxyartemisin   1α-hydroxydeoxyartemisinin, deoxyartemisinin, and 3α-hydroxydeoxyartemisinin.
2. **Phytochemicals:** Taxol.
   - ➤ **Origin and Bioactivity:** Taxusbaccata Anti-tumor.
   - ➤ **Microbial Catalyst:** Nocardioidesalbus.

> **Transformed Products:** 13-deacylpaclitaxel.
3. **Phytochemicals:** Ursolic acid (triterpenoid).
    > **Origin and Bioactivity:** Anti-cancer.
    > **Microbial Catalyst:** Nocardia sp. NRRL. 5646.
    > **Transformed Products:** Oleanolic acid methyl ester, oleanolic acid, and ursolic acid methyl ester.
4. **Phytochemicals:** Progesterone.
    > **Origin and Bioactivity:** Pregnant mares / Hog ovaries Sex hormone.
    > **Microbial Catalyst:** *Thamnostylum pyriforme* ATCC 8992 and *Mucor griseo-cyanus* ATCC 1207.
    > **Transformed Products:** 14a-OH progesterone, 9a-OH progesterone, 14a-OH progesterone, 7a, 14a, di-OH progesterone, and 6b, 14a-OH progesterone.
5. **Phytochemicals:** Asiatic acid.
    > **Origin and Bioactivity:** Antitumor and anti-cancer.
    > **Microbial Catalyst:** *Alternaria longipesas* 3.2875.
    > **Transformed Products:** 2a, 3b, 23,30-tetrahydroxyurs-12-ene-28-oic acid, 2a, 3b, 22b, 23-tetrahydroxyurs-12-ene-28-oic acid and 2a, 3b, 22b, 23,30-pentahydroxyurs-12-ene-28-oic acid.
6. **Phytochemicals:** Caffeine.
    > **Origin and Bioactivity:** Coffea arabica CNS stimulant.
    > **Microbial Catalyst:** *Pseudomonas putida* and *Penicillium roqueforti*.
    > **Transformed Products:** Theobromine 3-desmethyl caffeine and Biodegradation (Imidazole ring breakage.

4.5 CONCLUSION

Microbial transformations of natural compounds picked up their significance with the advancement of steroid drugs where such procedures play a vital work. Biotransformation is the premise of life. Microorganisms have been generally related for steroid biotransformation to get ready specific derivatives, the creation of which is troublesome by conventional manufactured strategies. The acknowledgment of microbial biotransformation as a significant assembling device has expanded in pharmaceuticals. The aim of the chapter is to present the generation of natural pharmaceuticals based on the microbial transformation of herbal constituents. Depicted are some effective uses of microbial transformation methods for the generation of steroid drugs or potentially their significant phytoconstituents as well as microbial modifications of terpenes, alkaloids, flavonoids bearing subsidiaries with improved biological activities.

KEYWORDS

- **herbal constituents**
- **microbial transformation**
- **natural pharmaceuticals**
- **phytoconstituents**
- **polychlorinated biphenyls**
- **steroid drugs**

REFERENCES

1. Asha, S., & Vidyavathi, M., (2009). Cunninghamella—a microbial model for drug metabolism studies: A review. *Biotechnol. Adv., 27*(1), 16–29.
2. Pervaiz, I., Ahmad, S., Madni, M. A., Ahmad, H., & Khaliq, F. H., (2013). Microbial biotransformation: A tool for drug designing (Review). *Applied Biochemistry and Microbiology, 49*(5), 435–449.
3. Cresnar, B., & Petric, S., (2011). Cytochrome P450 enzymes in the fungal kingdom. *Biochim. Biophys. Acta, 1814*(1), 29–35.
4. Bianchini, L. F., Arruda, M. F. C., Vieira, S. R., Campelo, P. M. S., Grégio, A. M. T., et al., (2015). Microbial biotransformation to obtain new antifungals. *Front Microbiol., 6*, 1433.
5. Timbrell, J. A., & Marrs, T. C. (2009). *Biotransformation of Xenobiotics. General and Applied Toxicology.*
6. Jones, A. L., & Fawcett, D. W., (1966). Hypertrophy of the granular endoplasmic reticulum in hamster liver induced by phenobarbital (with a review on the functions of this organelle in liver). *J. Histochem. Cytochem., 14*(3), 215–232.
7. Karigar, C. S., & Rao, S. S., (2011). Role of microbial enzymes in the bioremediation of pollutants: A review. *Enzyme. Res.,* p. 11.
8. Parkinson, A., (2001). Biotransformation of xenobiotics. In: Klaassen, C. D., (ed.), *Toxicology: The Basic Science of Poisons* (pp. 133–224). Access Pharmacy.
9. Kebamo, S., Tesema, S., & Geleta, B., (2015). The role of biotransformation in drug discovery and development. *J. Drug Metab. Toxicol., 6*, 196.

10. Huttel, W., & Hoffmeister, D., (2010). Fungal biotransformations in pharmaceutical sciences. In: Hofrichter, M., (ed.), *Industrial Applications, the Mycota X, 2* (pp. 293–317). Springer, Berlin, Germany.

11. Lilly, M. D., (1994). Advances in biotransformation processes. *Chem. Eng. Sci., 49*(2), 151–159.

12. Walker, J. M., & Cox, M. (1999). *The Language of Biotechnology: A Dictionary of Terms* (2nd edn.). USA: ACS Professional Reference Book, ACS; 199.

13. Collins, A. M., & Kennedy, M. J., (1999). Biotransformations and bioconversions in New Zealand: Past endeavors and future potential. *Austral. Biotechnol., 9*(2), 86–94.

14. Leresche, J. E., & Meyer, H. P., (2006). Chemocatalysis and biocatalysis (biotransformation): Some thoughts of a chemist and a biotechnologist. *Org. Proc. Res. Dev., 10*, 572–580.

15. Tang, F. H., Zhao, Y. J., & Tang, A. K., (2005). Presence of ectoparasitic trichodinids (Ciliophora, Oligohymenophorea, Peritrichida) on the gills of cultured fresh water fish, *Carassius auratus* in Chongqing, China, with the description of a new species of the genus. *Trichodina Acta Zootaxon Sin., 30*, 35–40.

16. Gershwin, L., (2006). Nematocysts of the cubozoa. *Zootaxa, 1232*, 1–57.

17. Hotta, Y., Ezaki, S., Atomi, H., & Imanaka, T., (2002). Extremely stable and versatile carboxylesterase from hyperthermophilic. *Appl. Environ. Microbiol., 68*(8), 3925–3931.

18. Chibata, I., & Wingard, L. B., (1983). Applied biochemistry and bioengineering. *Enzyme Technology, 1*, 1–355.

19. Foody, P., (1984). *Method for Increasing the Accessibility of Cellulose in Lignocellulosic Materials, Particularly Hardwoods Agricultural Residues and the Like*. USA.

20. Stankiewicz, A. I., & Jacob, A. M., (2000). Process intensification: Transforming chemical engineering. *Chem. Eng. Prog., 96*, 22–34.

21. Sultan, A., & RaufRaza, A., (2015). Steroids: A diverse class of secondary metabolites. *Med. Chem., 5*, 310–317.

22. Hegazy, M. E., Mohamed, T. A., ElShamy, A. I., Mohamed, A. E., Mahalel, U. A., et al., (2015). Microbial biotransformation as a tool for drug development based on natural products from mevalonic acid pathway: A review. *J. Adv. Res., 6*(1), 17–33.

23. Baiping, M., Feng, B., Huang, H., & Cong, Y., (2010). Biotransformation of Chinese herbs and their ingredients. *World Science and Technology, 12*, 150–154. doi: http://dx.doi.org/10.1016/S1876-3553(11)60012-4 (accessed on 16 May 2020).

24. Macedo, J. A., Battestin, V., Ribeiro, M. L., & Macedo, G. A., (2011). Increasing the antioxidant power of tea extracts by biotransformation of polyphenols. *Food Chemistry, 126*, 491–497.

25. Chen, Zhouzhou, Jian, L., Huimin, L., Lei, S., Jihong, Q., Xing, F., Xiaojing, D., & Daijie, C., (2013). Biotransformation of 14-deoxy-14-methylene triptolide into a novel hydroxylation product by *Neurospora crassa*. *Journal of Bioscience and Bioengineering, 116*, 199–202.

26. Alfarra, H. Y., Hasali, N. H. M., & Omar, M. N., (2013). A lignolytic fungi with laccase activity isolated from the Malaysian local environment for phytochemical transformation purposes. *International Research Journal of Biological Sciences, 2*, 51–54.

27. Takahashi, J. A., Gomes, D. C., Lyra, F. H., Dos, S. G. F., & Martins, L. R., (2014). The remarkable structural diversity achieved in ent-kaurane diterpenes by fungal biotransformations. *Molecules, 19*, 1856–1886.

28. Xia, Y. G., Yang, B. Y., & Kuang, H. X., (2015). Schisandraceae triterpenoids: A review. *Phytochem. Rev., 14*, 155–187.

29. Parikh, N. R., Mandal, A., Bhatia, D., Siveen, K. S., Sethi, G., et al., (2014). Oleanane triterpenoids in the prevention and therapy of breast cancer: Current evidence and future perspectives. *Phytochem. Rev., 13*, 793–810.

30. Feng, X., Zou, Z. M., Chu, Z. Y., & Sun, D. A., (2011). Biotransformation of taxanes. *Chinese Journal of Natural Medicines, 9*, 466–472.

31. Chen, T. S., Li, X., Bollag, D., Liu, Y. C., & Chang, C. J., (2001). Biotransformation of taxol. *Tetrahedron Letters, 42*, 3787–3789.

32. Shibuya, H., Kitamura, C., Maehara, S., Nagahata, M., Winarno, H., et al., (2003). Transformation of cinchona alkaloids into 1-N-oxide derivatives by endophytic Xylaria sp. isolated from cinchona pubescens. *Chem. Pharm. Bull. (Tokyo), 51*, 71–74.

33. Chen, N., Zhang, J., Liu, J., & Yu, B., (2010). Highly efficient and region-selective glucosylation of 25(S) ruscogenin by *Gliocladium deliquescens* NRRL 1085. *Chin. J. Chem., 28*, 439–442.

34. Falcone, F. M. L., Rius, S. P., & Casati, P., (2012). Flavonoids: Biosynthesis, biological functions, and biotechnological applications. *Front Plant. Sci., 3*, 222.

35. Ren, W., Qiao, Z., Wang, H., Zhu, L., & Zhang, L., (2003). Flavonoids: Promising anticancer agents. *Med. Res. Rev., 23*, 519–534.

36. Sultan, A., & RaufRaza, A., (2015). Steroids: A diverse class of secondary metabolites. *Med. Chem., 5*, 310–317.

37. Okafor, N., (2007). *Modern Industrial Microbiology and Biotechnology*. New Hampshire: Science Publishers.

38. Gupta, A., Singh, V. K., Qazi, G. N., & Kumar, A., (2001). *Gluconobacter oxydans*: Its biotechnological applications. *J. Mol. Microbiol. Biotechnol., 3*(3), 445–456.

39. Boopathy, R., (2000). Factors limiting bioremediation technologies. *Bioresour. Technol., 74*(1), 63–67.

40. Mathew, H., (2012). Refined politics: Petroleum products, neoliberalism, and the ecology of entrepreneurial life. *Journal of American Studies, 46*(2), 295–312.

CHAPTER 5

Actinidia deliciosa: A Nature's Boon to Modern Pharmacotherapeutics

VIVEK KUMAR RAMAN,[1] SUNIL KUMAR CHAUHAN,[1] and ARIJIT CHAUDHURI[2]

[1]Manav Bharti University, Solan – 173229, Himachal Pradesh, India, E-mail: vivekraman72@gmail.com (V. K. Raman)

[2]U.S. Ostwal Institute of Pharmacy, Mangalwad – 312024, Rajasthan, India

ABSTRACT

Actinidia deliciosa is also known as Green Kiwi, kiwi fruit, Chinese gooseberry, yang tao, etc. which belongs to the sub-family of genus *Actinidia.* The book chapter highlighted anti-hypertensive, anti-diabetic, anti-carcinogenic, anti-fungal, hepatoprotective, anti-asthma, anti-platelet, anti-nociceptive, anti-retroviral, etc. of Kiwi fruit. The methods and modes of production, storage, side effects, and traditional uses of Kiwi fruits are discussed in this chapter.

5.1   INTRODUCTION

Actinidia deliciosa is also known to be Green Kiwi, kiwi fruit, Chinese gooseberry, yang tao, etc. which belongs to the sub-family of genus *Actinidia.* It is distributed throughout the world, especially in eastern Asia [1]. It is a commercial crop in several countries, mainly in Italy, China, and New Zealand. With advanced research and development support, its cultivation in India has been expanded in the mid-hills of Jammu Kashmir, Himachal Pradesh, and Arunachal Pradesh [2]. China is the native origin of Kiwi fruit. The genus name Actinidia refers to the Greek word *aktinos* (rays) which refers to the styles of the female flower, which radiate from the center and resemble the spokes of a wheel. In India, the cultivated area of this fruit is very less. Kiwi fruit contains an actinidin enzyme, which is often considered as an allergen. Furthermore, it is a unique source of various nutrients and phytochemicals to develop various medicines against different diseases. Healthful attributes of Kiwi due to the presence of vitamins, folic acid, carbohydrates, minerals, amino acid, saponins, tannins, flavonoids, and steroids [3–5]. It also contains kissper peptide, which is responsible for the anti-inflammatory and antioxidant properties. Kiwi contains various phytochemical properties such as antihypertensive, antidiabetic, anticarcinogenic, antifungal, hepatoprotective, anti-asthma, anti-platelet, anti-nociceptive, anti-retroviral, etc. Thus, Kiwi is cultivated for its nutritional benefits and useful medicinal properties [6–22]. *A. deliciosa* is economically the most important crop and its total production is about 1.9 million tons per year. However, at the international level, it is considered a minor crop that represents about 0.2–0.3% of total fresh fruit production [23, 24]. The scientific classification of the Kiwi plant is given in Table 5.1.

TABLE 5.1   Scientific Classification of Kiwi Plant

Classification	Name
Kingdom	Plantae
Division	Magnoliophyta
Class	Magnoliopsida
Subclass	Magnoliidae
Order	Ericales
Superorder	Asteranae
Family	Actinidiaceae
Genus	Actinidia
Species	deliciosa

5.2   DESCRIPTION

A. deliciosa is a woody, vigorous, climbing shrub reaching approximately 9 m. Its leaves are long-petioled, alternate, oval to nearly circular, deciduous, heart-shaped at the bottom, and have 8–13 cm length. Young leaves are coated with red-colored hairs; mature leaves are dark-green and smooth-faced on the upper side, and downy-white with distinguished, light-colored veins beneath. The flowers are aromatic, dioecious or sexual, borne individually or in threes within the leaf axils are 4–6 petalled, white initially, changing to buff-yellow, 3–6 cm wide and each sex contains central tufts of several stamens, though the female flowers with no viable pollen. The oblongated fruits are up to 7–8 cm long. The reddish-brown skin of the fruits is fully coated with stiff, short, brown hairs. The flesh is firm until completely ripened; it is bright, juicy, and luscious. The color of the flesh is bright green, or sometimes brownish, yellow or off-white. The flavor is sour to quite acid; the flavor is suggested to be similar to that of the gooseberry or strawberry [25].

5.3 TRADITIONAL USES

In the roots of *A. deliciosa*, various folk remedies for adult diseases, such as anti-pyorrheal, anti-hepatotoxic, and gingival inflammation has been observed [27]. The genus *Actinidia* is used in Chinese folk medicine system for the treatments of diseases such as hepatitis, edema, rheumatoid arthritis, gastric cancer, breast cancer, etc. on a wide scale [28]. Kiwi fruit is distributed in west China and shown to possess anti-tumor and protective effects on acute hepatic injury in biological arrays [29]. It was also reported to have gentle laxative effects, because of its significant levels of dietary fibers [30]. *A. deliciosa* components, presumably involving vitamin E and omega-3 fatty acids from its edible seeds, have potential properties of a natural blood thinner [31]. *A. deliciosa* may be a natural supply of carotenoids like pro-vitamin A-beta-carotene [32], lutein, and zeaxanthin [33]. The fruits, roots, and stems have diuretic, pyretic, and sedative properties. These are used to treat rheumatoid arthralgia disease, cancers of the liver and esophagus, and stone removal in the urinary tract. Usually, Kiwi fruit is eaten fresh, but it may also be taken in desserts, beverages, and also as a flavoring. The fruits are very high in vitamin C. It also contains vitamin A and vitamin E; also, it contains a sufficient amount of potassium [34]. Kiwi fruit is a rich source of flavonoid antioxidants (AOs). The seed oil of *A. deliciosa* contains approximately 60–64% alpha-linolenic acid, which is an omega-3 fatty acid [35]. Inositol found in *A. deliciosa* works as a precursor of an intracellular second messenger system. It is beneficial to treat depression [36]. Due to the presence of actinidin enzyme, it has application as a meat tenderizer. It can be used for making wine, jams, and cocktails, and is used in the preparation of seafood, chicken, and ham.

5.4   ACTIVE CONSTITUENTS OF *ACTINIDIA DELICIOSA*

The fruiting body of *A. deliciosa* contains a broad range of active constituents which include amino acids (histidine, arginine, tyrosine, valine, and phenylalanine), vitamins (vitamin B_1 (thiamine), vitamin B_2 (riboflavin), vitamin B_3 (niacin), vitamin B_6 (pyridoxine), vitamin B_9 (folate), vitamin C, vitamin E, and vitamin K), anthocyanins (carotenoids, beta-carotene, and lutein), organic acids (citric acid, quinic acid, and maleic acid), tannins, etc. The other parts of kiwi plant also contain various active constituents like phenolic acids (vanillic acid, hydroxyl cinnamic acid, and caffeic acid), coumarins (umbelliferon and fraxetin), steroids (sitosterol), sesquiterpenoids (alpha-farnesene and germacrene D), carbohydrates (starch, cellulose, pectin, and sugars), minerals (Mg, P, Mn, K, Na, and Zn), protein (actinidin), flavonoids (quercetin and kaempferol), organic acids (citric acid and quinic acid), etc. [37, 38].

5.5   PHARMACOLOGICAL PROPERTIES OF *ACTINIDIA DELICIOSA*

The kiwi fruit possesses various pharmacological properties such as cytotoxic, anti-hypertensive, anti-hypercholesterolemia, anti-oxidant, anti-tumor, anti-proliferative, anti-carcinogenic, anti-inflammatory, anti-microbial, anti-spermatogenesis, anti-constipation, anti-fungal, hepatoprotective, anti-asthma, anti-oxidative stress, anti-platelet, fibrinolytic, anti-thrombin, anti-atherosclerotic, dermatologic, etc. [6–22].

5.5.1   *NUTRITIONAL IMPORTANCE*

Kiwi fruits are full of nutrients and according to depth analysis and research over the last decade on the health, benefits have shown that regular consumption of *A. deliciosa* demonstrates improvements not only in nutritional status, yet also in offer benefits to immune, digestive, and metabolic health [39]. Kiwi fruit contains a high amount of vitamin C and also contain other nutrients such as potassium, dietary fiber, vitamin E, and folate with nutritionally relevant levels. It also contains various bioactive components that include a wide range of phytonutrients, AOs, and enzymes that help in providing functional and metabolic benefits. In digestive health, the contribution of kiwi is attracting specifically considered attributable to a developing assortment of proof from human intercession studies. There are numerous conceivable mechanisms of action that are probably going to act together including the fiber content and type, the presence of actinidin (a natural proteolytic enzyme that breaks down protein and facilitates gastric and ideal digestion [40, 41]), and different phytochemicals that can stimulate motility [42].

5.5.2 ANTI-CANCER PROPERTIES

An anti-mutagenic component found in *A. deliciosa* helps to prevent mutations of genes which can originate the cancer process. Glutathione is also present in *A. deliciosa* which may help in the reduction process. During smoking or grilling of foods, carcinogenic nitrates are formed, when these particles ingested in the body a process called nitrosation happens. When it occurs free radicals of nitrosamines are formed that may stimulate the development of the gastric or different types of cancers. The amino acid arginine, present in *A. deliciosa* was investigated by the cardiologists to improve post-angioplasty blood flow and avert the arrangement or reconstruction of plaque inside the arteries. This fruit is on the 4th-position in the highest natural antioxidant potential after red fruits containing higher levels of beta-carotene. Lutein, a phytochemical found in *A. deliciosa* may help in the prevention of lung and prostate cancer [43].

5.5.3 CARDIOVASCULAR DISEASE

Platelet hyperactivity is one of the most significant hazard factors in charge of the frequency of cardiovascular infection. Numerous nutritive and non-nutritive compounds present in the fruits and vegetables may affect platelet function in many different ways. A recent discovery of anti-platelet factors in *A. deliciosa* provided a new dietary source as a preventive restorative technique to alter platelet action. Kiwi fruits are rich in glutamate and arginine. Arginine may help to promote an increase in arteriolar dilation, may work as a vasodilator, and can improve bloodstream important for heart wellbeing. Fruits thin up blood reduce clotting by an average of 15–18% and lower your fat in the blood by an average of 13–15% [43].

5.5.4   ANTIDIABETIC ACTIVITY

Inositol, a natural sugar alcohol found in *A. deliciosa* helps in regulating diabetes. Inositol supplements can improve nerve conduction velocity in diabetic neuropathy. It plays a noteworthy job in intracellular reactions to hormones and neurotransmitters. Inositol acts as a second messenger in the cell signaling process [43].

5.5.5   ANTI-INFLAMMATORY ACTIVITY

There is much research enthusiasm for distinguishing new dietary anti-inflammatory agents that can prevent chronic disease development and other degenerative processes. *A. deliciosa* and its constituents have been one of the topics of such investigations. Few animal studies evaluating the anti-inflammatory effects of *A. deliciosa* have been reported [44].

5.5.6   DERMATOLOGICAL ACTIVITY-BURN TREATMENT

Two recent rat studies gave a fascinating report which shows that dressing prepared from slices of fresh *A. deliciosa* leads to promote healing of acute burn wounds [45]. Two rat groups were taken. One group was treated with silver sulfadiazine cream (ointment used in topical burn management) and the other group was treated with the dressing prepared from slices of fresh *A. deliciosa*. The wound healed rapidly and dry scars detached early in the *A. deliciosa* treated group. In addition to it, anti-bacterial, and angiogenic actions were observed [46]. It was also reported that there was no positive culture medium for *Streptococcus pseudomonas* or *Staphylococcus* in the *A. deliciosa* treated rodents group. A proposed system for the improved injury debridement included the beneficial proteolytic activity of actinidin and other degradative enzymes known to be available in *A. deliciosa*. The components responsible for the anti-inflammatory, angiogenic, and antimicrobial actions were not confirmed. The active constituents of the *A. deliciosa* that are responsible for the various valuable results need to be identified [44].

5.6   SIDE EFFECTS ASSOCIATED WITH KIWI FRUIT

The most common side effect is of kiwi is an allergy which may be characterized by local mouth irritation to anaphylaxis. Acute pancreatitis has conjointly been reported. Because of high levels of vitamin C and E and potassium, it may be capable of altering triglycerides (TG) levels [26].

5.7   PRODUCTION

The most broadly planted kiwi fruit cultivar is the fluffy kiwi fruit *A. deliciosa*- 'Hayward.' 'Hayward' represents about a half portion of kiwi cultivation all of the worlds. It also represents about 90–95% of the kiwi fruit traded internationally [47]. The fluffy kiwi fruit *A. deliciosa* is economically the most important crop and its total production is about 1.6–2 million tons per year. However, in the international market, total kiwi fruit production is only about 0.2–0.3% of the total fresh fruit production [48]. In terms of "marketable gross production," it is the 6th most valuable fruit crop. China is the largest producer of kiwi fruit after that Italy, New Zealand, Chile, and Greece were followed.

5.8   STORAGE

Kiwi fruits can be kept for 4 to 6 months when cultivated at correct maturity and at optimal storage, conditions (gas composition, temperature, ethylene concentration, and relative humidity). Kiwi fruits that are proposed for sale within 2–4 months are stored under normal atmospheric conditions [49], while fruits that are proposed to store for a longer period are stored under artificial atmospheric conditions. Earlier, MAP with 3% oxygen and 3% carbon dioxide was used, but later oxygen concentrations were reduced and carbon dioxide concentrations were increased. The reason was higher carbon dioxide concentrations reduce fruit ripening because of reduced fruit respiration [50]. Nowadays, fruits are stored under 4.5–5%

carbon dioxide and 1.8–2% oxygen in CA stores. ULO (ultra-low oxygen) technique has also been used, but it did not seem to be suitable for kiwi fruit as it develops of off-flavors and hence reduced storage life. Within cold storages, the relative humidity is maintained above 94–95% that helps to reduce weight loss by 2.5–7% [51]. Kiwi fruits are sensitive to ethylene and concentrations as low as 0.1–1 ppm can induce softening [52]. In cold storages, ethylene concentrations are kept below 0.05 ppm to prevent softening. Different preservation techniques including chemical dipping, cold storages, artificial atmospheric condition, and edible coatings have been used to increase the shelf-life and preserve the nutritional value of fresh-cut kiwi fruits.

5.9 CONCLUSION

Kiwi fruit has been consumed by humans since ancient times. Scientific researchers have proved that the daily consumption of Kiwi fruits in daily life can help in reducing various diseases. It is one of the most popular delicious fruits, which possesses lots of medicinal properties. In this book chapter, a collection of all the necessary information on Kiwi fruit is collected, which may help the researchers and students to achieve quality knowledge. Kiwi fruit was originated from China. It is a package full of bioactive compounds, nutrients, and minerals, which make it a very special dietary supplement. It is useful in the successful management of various diseases such as HIV, inflammation, hypertension, cancer, asthma, and diabetes. From ancient times, it has been used as a diuretic, mild laxative, and anti-hepatotoxic. It showed an excellent anti-oxidant property. Clinical trials need to be carried out to exploit the therapeutic utility of Kiwi in combating various diseases.

KEYWORDS

- *Actinidia deliciosa*
- **anti-cancer properties**
- **anti-inflammatory agents**
- **anti-oxidant property**

- **chronic disease**
- **kiwi fruit**
- ***Streptococcus pseudomonas***
- **ultra-low oxygen**

REFERENCES

1. Ferguson, A. R., (1990). Botanical nomenclature: *Actinidia chinensis*, *Actinidia deliciosa*, and *Actinidia setosa*. In: *Kiwifruit: Science and Management* (pp. 36–57). Warrington I.J., Weston G.C.
2. Atkinson, R. G., & Macrae, E. A., (2007). Kiwifruit. In: Pua, E. C., & Davey, M. R., (eds.), *Transgenic Crops V* (Vol. 60, pp. 329–346). Springer Berlin Heiderberg.
3. Parameswaran, I., & Murthi, V. K., (2014). Comparative study on physico and phyto-chemical analysis of *Persea americana* and *Actinidia deliciosa*. *International Journal of Scientific and Research Publications, 4*(5), 1–5.
4. Rush, E. C., Patel, M., Plank, L. D., & Ferguson, L. R., (2003). Kiwifruit promotes laxation in the elderly. *Asia Pacific Journal of Clinical Nutrition, 11*(2), 164–168.
5. Teng, K., Ruan, H. S., & Zhang, H. F., (2013). Flavonoid and saponin rich fractions of kiwi roots (*Actinidia arguta* (Sieb.et Zucc.) Planch) with antinociceptive and anti-inflammatory effects. *African Journal of Pharmacy and Pharmacology, 7*(35), 2445–2451.
6. Al-Naimy, E. H., Al-Lihaibi, R. K., Majeed, S. M., & Al-Ani, R. S., (2012). Antibacterial and cytotoxic effects of the kiwi fruit and Pomegranate active compounds on tumor cell line (L20B, RD). *Iraqi Journal of Agriculture Science, 43*(1), 157–167.
7. Ah Jung, K., Song, T. C., Han, D., Kim, I. H., Kim, Y. E., & Lee, C. H., (2005). Cardiovascular protective properties of kiwi fruit extracts *in vitro*. *Biology and Pharmacology Bulletin, 28*(9), 1782–1785.
8. Zuo, L. L., Wang, Z. Y., Fan, Z. L., Tian, S. Q., & Liu, J. R., (2012). Evaluation of anti-oxidant and anti-proliferative properties of three actinidia (*Actinidia kolomikta*, *Actinidia arguta*, and *Actinidia chinenesis*) Extracts *in vitro*. *International Journal of Molecular Sciences, 13*(5), 5506–5518.
9. Yu Ku, C., Wang, Y. R., Yuan, L. H., Chun, L. S., & Yaw, L. J., (2015). Corosolic acid inhibits hepatocellular carcinoma cell migration by targeting theVEGFR2/Src/FAK pathway. *PLOS One, 10*(5), e0126725.
10. Sadek, M. A., Aref, M. L., Khalil, F. A., Barakat, L. A. A., Ali, N. H., & Saliman, B. S. M., (2012). Impact of *Actinidia deliciosa* (Kiwi fruit) consumption on oxidative stress status in carcinogenesis. *African Journal of Biological Sciences, 8*(1), 117–127.
11. Ciacci, C., Russo, I., Bucci, L. P., Pellegrini, L., Giangrieco, I., Tamburrini, M., & Ciardiello, M. A., (2013). The kiwi fruit peptide kisper displays anti-inflammatory and anti-oxidant effects in *in-vitro* and *ex-vivo* human intestinal models. *The Journal of Clinical and Experimental Immunology, 175*(3), 476–484.

12. Mishra, N., Dubey, A., Singh, N., & Gupta, P., (2010). Antimicrobial potential of vitamin rich fruits. *International Journal of Applied Biology and Pharmaceutical Technology, 1*(3), 915–920.

13. Dehghani, F., Khozani, T. T., Panjehshanin, M. R., & Panahi, Z., (2006). Toxic effects of hydroalcoholic extract of Kiwi (*Actinidia chinensis*) on histological structure of the male Sprague dawley rat reproductive tissue. *Iranian Journal of Science and Technology, 30*, 19–25.

14. Maleleo, D., Gallucci, E., Notarochille, G., Sblano, C., Schettino, A., & Micelli, S., (2012). Studies on the effects of salts on the channel activity of kissper, a Kiwifruit peptide. *The Open Nutraceuticals Journal, 5*, 136–145.

15. Wang, H., & Ng, T. B., (2002). Isolation of an anti-fungal thaumatin-like protein from Kiwifruits. *The Journal of Phytochemistry, 61*(1), 1–6.

16. Amer, M. A., Eid, J. I., & Hamad, S. R., (2014). Evaluation of gastric and hepatic protective effect of kiwifruit extract on toxicity of Indomethacin in Swiss albino mice using histological studies. *International Journal of Science and Research, 3*(7), 1631–1641.

17. Forastiere, F., Pistell, R., Sestini, P., Fortes, C., Renzoni, E., Rusconi, F., Dell'orco, V., & Ciccone, G., (2000). Bisanti L and Sidria collaborative group, consumption of the fresh fruits rich in vitamin C and wheezing symptoms in children. *Thorax, 55*, 283–288.

18. Brevik, A., Gaivao, I., Medin, T., Jorgenesen, A., Piasek, A., Elilason, J., Karlsen, A., et al., (2011). Supplementation of western diet with golden Kiwi fruit (*Actinidia chinensis* var. 'Hort 16 A:) effects of biomarkers of oxidation damage and anti-oxidation protection. *Nutrition Journal, 10*, 1–9.

19. Teng, K., Ruan, H. S., & Zhang, H. F., (2013). Flavanoids and saponin rich fractions of Kiwi roots (*Actinidia arguta* (Sieb.et Zucc.) Planch) with anti-nociceptive and anti-inflammatory effects. *African Journal of Pharmacy and Pharmacology, 7*(35), 2445–2451.

20. Urrutia, C. T., Guziman, L., Hirschmann, G. S., Carrasco, M., Alarcon, M., Astudillo, L., Gutierrez, M., Carrasco, G., Yuri, J. A., Aranda, E., & Palomo, I., (2011). *Antiplatelet, Anticoagulant and Fibrinolytic Activity In Vitro Extracts from Selected Fruits and Vegetables* (Vol. 22, pp. 197–205). Lippincott Williams and Wilkins.

21. Shehata, M. M. S. M., & Saltan, S. S. A., (2013). Effects of bioactive component of Kiwifruit and avocado (fruit and seed) on hypercholestertolemic rats. *World Journal of Dairy and Food Sciences, 8*(1), 82–93.

22. Deters, A. M., Schroder, K. R., & Hensel, A., (2005). Kiwifruit (*Actinidia deliciosa* L.) polysaccharides exert stimulating effects to on cell proliferation via enhanced growth factor receptors, energy production, and collagen synthesis of human keratinocytes, fibroblasts, and skin equivalents. *Journal of Cell Physiology, 202*(3), 717–722.

23. Belrose Inc., (2007). *World Kiwifruit Review-I.* Belrose Inc, Pullman, WA.

24. Belrose Inc., (2008). *World Kiwifruit Review-II.* Belrose Inc, Pullman, WA.

25. Morton, J., (1987). Kiwifruit. In: *Fruits of Warm Climates* (pp. 293–300). Julia F. Morton, Miami, FL.

26. Harsh, C., Milind, P., & Monu, Y., (2016). *Medicinal Potential and Phytopharmacology of Actnidia Deleciosa, 6*(1), 20–25.

27. *Chinese Traditional Medicine Glossary*, (1977). Shanghai Science and Technology Publishing Co., p. 2211.

28. Jiangsu New Medical College, (1977). *Dictionary of Chinese Herbal Medicines* (p. 2210). Shanghai: Shanghai Science and Technology Press.

29. Zhong, Z. G., Zhang, F. F., Zhen, H. S., et al., (2004). Experimental study on the anti-tumor effects of extracts from roots of *Acitinidia delicosa* in carcinoma cell lines [J]. *Chin. Arch. Trad. Chin. Med.*, pp. 1705–1707.

30. Rush, et al., (2002–2006). *Actinidia deliciosa* promotes laxation in the elderly. *Asia Pacific Journal of Clinical Nutrition*.

31. Duttaroy, A. K., & Jørgensen, A., (2004). Effects of kiwi fruit consumption on platelet aggregation and plasma lipids in healthy human volunteers. *Platelets*, pp. 287–292.

32. Kim, M., Kim, S. C., Song, K. J., Kim, H. B., Kim, I. J., Song, E. Y., & Chun, S. J., (2010). Transformation of carotenoid biosynthetic genes using a micro-cross section method in *Actinidia deliciosa* (*Actinidia deliciosa* cv. Hayward). *Plant. Cell Rep.*, pp. 1339–1349.

33. Sommerburg, O., Keunen, J. E., Bird, A. C., & Van, K. F. J., (1998). Fruits and vegetables that are sources for lutein and zeaxanthin: The macular pigment in human eyes. *Br. J. Ophthalmol.*, pp. 82–89.

34. Ferguson, A. R., (1990). The genus actinidia. In: Warrington, I. J., & Weston, G. C., (eds.), *Actinidia Deliciosa: Science and Management*. Ray Richards Pub., Auckland, New Zealand.

35. Selman, J. D., (1983). The vitamin C content of kiwi fruit (*Actinidia deliciosa*, variety). *Food Chem.*, pp. 63–75.

36. Lal, S., Ahmed, N., Singh, S. R., Singh, D. B., Mir, J. I., & Kumar, R., (2010). *Actinidia Deliciosa: Miracle Berry* (pp. 52–55). Science Reporter.

37. Parameswaran, I., & Murthi, V. K., (2014). Comparative study on physico and phyto-chemical analysis of *Persea americana* and *Actinidia deliciosa*. *International Journal of Scientific and Research Publication, 4*(5), 1–5.

38. Motohashi, N., Shirataki, Y., Kawase, M., Tani, S., Sakagmi, H., Satoh, K., Kurihara, T., Nakashima, H., Wolfard, K., Miskolci, C., & Molnar, J., (2001). Biological activity of Kiwifruit peel extract. *Phytotherapy Research, 15*(4), 337–343.

39. Boeing, H., Bechthold, A., Bub, A., Ellinger, S., Haller, D., Kroke, A., Leschik-Bonnet, E., Müller, M. J., Oberritter, H., Schulze, M., Stehle, P., & Watzl, B., (2012). Critical review: Vegetables and fruit in the prevention of chronic diseases. *Eur. J. Nutr., 51*, 637–663.

40. Kaur, L., Rutherfurd, S. M., Moughan, P. J., Drummond, L., & Boland, M. J., (2010). Actinidin enhances protein digestion in the small intestine as assessed using an *in vitro* digestion model. *J. Agric. Food Chem., 58*(8), 5074–5080.

41. Kaur, L., Rutherfurd, S. M., Moughan, P. J., Drummond, L., & Boland, M. J., (2010). Actinidin enhances gastric protein digestion as assessed using an *in vitro* gastric digestion model. *J. Agric. Food Chem., 58*(8), 5068–5073.

42. Ciardiello, M. A., Meleleo, D., Saviano, G., Crescenzo, R., Carratore, V., Camardella, L., Gallucci, E., Micelli, S., Tancredi, T., Picone, D., & Tamburrini, M., (2008). Kissper, a kiwi fruit peptide with channel-like activity: Structural and functional features. *J. Pept. Sci., 14*(6), 742–754.

43. Lal, S., Ahmed, N., Singh, S. R., Singh, D. B., Mir, J. I., & Kumar, R., (2010). *Actinidia Deliciosa: Miracle Berry* (pp. 52–55). Science Reporter.
44. Keith, S., (2012). *Actinidia deliciosa*: Overview of potential health benefits. *Nutrition Today, 47*(3), 133–147.
45. Hafezi, F., Rad, H., Naghibzadeh, B., Nouhi, A., & Naghibzadeh, G., (2010). *Actinidia deliciosa*: A new drug for enzymatic debridement of acute burn wounds. *Burns, 36*, 352–355.
46. Mohajeri, G., Masoudpour, H., Heidarpour, M., et al., (2010). The effect of dressing with fresh *Actinidia deliciosa* on burn wound healing. *Surgery, 148*, 963–968.
47. Ferguson, A. R., & Seal, A. G., (2008). Kiwifruit. In: Jim, F., (ed.), *Temperate Fruit Crop Breeding, Hancock, Germplasm to Genomics*. East Lansing, MI, USA.
48. Guroo, I., Wani, S. A., Wani, S. M., Ahmad, M., Mir, S. A., & Masoodi, F. A., (2017). *A Review of Production and Processing of Kiwifruit, 8*(10), 1000699.
49. Brigati, S., & Donati, I., (2003). *Actinidia: Search Results for Their Applications in the Field of Conservation and Commercial Distribution* (pp. 277–290). Societa Orticola Italiana Verona, Italy.
50. Sozzi, A., Testoni, A., Youssef, J., Deluisa, A., & Nardin, C., (1980). *Preservation of Actinidia in Controlled Atmosphere* (Vol. 11, pp. 271–288). Annali Experimental Institute for the Technological Valorization of Agricultural Products, Italy.
51. Nardin, C., & Galliano, A., (1988). Technologies for refrigerant preservation of the product. *Proceedings on Actinidia* (pp. 135–150). Saluzzo.
52. Monzini, A., & Gorini, F., (1986). *Aspects and Problems of Preservation of Actinidia: Cultivation of Actinidia* (pp. 141–170). Italian Horticultural Society, Verona, Italy.

CHAPTER 6

Ocimum sanctum L. (Holy Basil or Tulsi): A Medicinally Significant Herb

MANIK DAS,[1,2] DEBARSHI KAR MAHAPATRA,[3] and KUNTAL MANNA[1]

[1]*Department of Pharmacy, Tripura University (A Central University), Suryamaninagar – 799022, Tripura, India*

[2]*Department of Pharmaceutical Chemistry, Srikrupa Institute of Pharmaceutical Sciences, Hyderabad – 502277, Telangana, India*

[3]*Department of Pharmaceutical Chemistry, Dadasaheb Balpande College of Pharmacy, Nagpur – 440037, Maharashtra, India*

ABSTRACT

Plant-based medicines or herbal drugs have an essential role in developing novel therapeutics for innumerable diseases. Traditional medicines (TMs) have a long history date back to thousands of years. Asian countries like India and China have a rich and diverse legacy of traditional systems of medicines like Ayurveda, Siddha, and Traditional Chinese Medicine. *Ocimum sanctum Linn* also is known as "Holy basil" or "*Tulsi*" is one of the indispensable medicinal plants (MPs) reported in TMs. It has been reported to possess diverse pharmacological activities like anti-microbial, immunomodulatory, anti-stress, anti-inflammatory, anti-ulcer, anti-diabetic, hepatoprotective, chemoprotective, anti-hyperlipidemic, cardio-protective, anti-oxidant, antitussive, radioprotective, memory enhancing, anti-arthritic, anti-fertility, anti-hypertensive, anti-coagulant, anti-cataract, anthelmintic, and anti-nociceptive activities. It has a safe record of human consumption for thousands of years. Hence, the plant has been extensively studied by several investigators and its active constituents are isolated and structure elucidations are done with modern analytical techniques.

Some of the individual components were comprehensively studied at the molecular level and the ongoing quest of researchers enabled us to know much about the chemical constitution of *O. sanctum Linn*. Therefore, in this book chapter, the summarization is done along with structures of phytochemicals of sacred basil with reported biological activities.

6.1 INTRODUCTION

From the beginning of civilization, plant-based medicines and medicinal plants (MPs) as a whole, in the form of tinctures and extracts were the companions of human society to counter countless diseases and disorders. This epitomizes that the plant kingdom holds an ample reservoir of phytoconstituents with therapeutic potential [1–5]. Quest of humans leads to the discovery of many active pharmaceutical ingredients which assisted the development of new chemical entities (NCEs) for various pathological states and also would serve the same in the future [4, 5]. One of the most resourceful MPs is *Ocimum sanctum.* Holy Basil or Tulsi, an Ayurvedic herb native to tropical regions of Asia, Africa, and Central and South America, scientifically or botanically named *O. sanctum* Linn. It has been found to have well-documented use in traditional medicine (TM) or folk medicine which encompasses medical aspects of traditional knowledge that developed over generations within various societies before the era of modern medicine. It is known for its wide spectrum biological activities since the last 2000 years and also considered as sacred or holy in many parts of the world especially in India. *Ocimum* sp. was found to possess anti-diabetic, wound healing, anti-oxidant, radiation protective, immunomodulatory, anti-fertility, anti-inflammatory, antimicrobial, anti-stress, and anti-cancer activities [6–11]. This chapter deals with the phytochemical details and pharmacological aspects of *O. sanctum* in the following sections.

6.2 TAXONOMICAL CLASSIFICATION OF HOLY BASIL OR TULSI

- ➢ **Kingdom:** Plantae.
- ➢ **Order:** Lamiales.
- ➢ **Family:** Lamiaceae.
- ➢ **Genus:** Ocimum.

> ➢ **Species:** *O. tenuiflorum/sanctum.*
> ➢ **Binomial name:** *Ocimum tenuiflorum* or *Ocimum sanctum L.*

6.3 NUTRACEUTICAL IMPORTANCE

Ocimum sanctum L. (Ocimum s.) is a well-established source of vitamins, minerals, fat, protein, polysaccharide, fiber, pigments, and mucilage. It contains calcium, phosphorus, zinc, iron, and copper which are minerals. It also contains Vitamin A, C, Niacin, Vit B_1, and Vit B_{12}. Hence, it is a valuable plant for food and nutraceutical industry [6–11].

6.4 TRADITIONAL AND AYURVEDIC IMPORTANCE

Tulsi got its name indexed in the ancient texts of Ayurveda including Charak Samhita, Susrut Samhita, and Rigveda (3500–1600 BCE) for the treatment of cough, respiratory disorders, poisoning, impotence, and arthritis. Various parts of *Ocimum sp.* such as leaves, roots, seeds, and also as a whole plant were documented for traditional uses for prevention and treatment of many illnesses and common ailments like the common cold, headache, cough, flu, earache, fever, colic pain, sore throat, bronchitis, asthma, hepatic diseases, malaria fever, as an antidote for snakebite and scorpion sting, flatulence, migraine headaches, fatigue, skin diseases, wound, insomnia, arthritis, digestive disorders, night blindness, diarrhea, and influenza. Tulsi leaves chewing may have a preventive effect on ulcers and infections of the mouth. Some ayurvedic formulations containing Tulsi as a major ingredient are anu taila (ayurvedic oil), marichyaadi taila, tulasiswarasadi taila, manasamitra vati (ayurvedic tablet), bilvadi gutika (Ayurvedic tablet), surasaadigana kwatha (coarse powder meant for decoction), ksudraadi kwatha, swasahara kashaya churna (Ayurvedic medicinal powders that are derived from plants and minerals), kustadi lepa (for topical applications), dasamoola grita (formulation with ghee), muktaadi mahanjana (medicated oil, mainly used in the treatment of eye disorders), cintamani rasa (Herbo-Metal preparation), sitakesari rasa, amavata rasayoga, and tulasi arka (preparation obtained by distillation), etc. These formulations are indicated in various disease conditions as per Ayurvedic formulary [12, 13].

6.5   CHEMICAL CONSTITUENTS OF *OCIMUM SANCTUM*

The phytochemical composition of Tulsi is highly multifaceted. Therefore, the isolation and identification of individual active constituents are relatively tedious and time-consuming. But with the advancement of modern pharmaceutical science, vast numbers of compounds present in Tulsi were isolated; structure elucidations were done and identified. Due to the presence of diverse phytoconstituents, pharmacology is also reasonably diverse. The nature of phytochemicals of *Ocimum sp.* and constituents reported are summarized in Table 6.1 [6–14].

TABLE 6.1   Phytochemicals of *Ocimum sanctum* L.

Nature of Chemical Constituents	Constituents Reported
Phenolics	Caffeic acid, chlorogenic acid, vanillic acid, ocimumnaphthanoic acid, menthylsalicylic glucoside, gallic acid, gallic acid methyl ester, gallic acid ethyl ester, protocatechuic acid, 4-hydroxybenzoic acid, vanillin, 4-hydroxybezaldehyde, and rosmarinic acid
Flavonoids and glycosides	Luteolin, isothymusin, cirsimartin, orientin, isoorientin, isovitexin, vicenin, apigenin, cirsimaritin, salvigenin, crisilineol, eupatorin, isothymusin, gardenin, flavone-7-O-glycosides, and luteolin-5-O-glucoside
Phenyl propanoids	Eugenol, ociglycoside or eugenyl-β-D-glucose, citrusin C, ferulaldehyde, bieugenol, and dehydrodieugenol
Neolignans	Tulsinol A, Tulsinol B, Tulsinol C, Tulsinol D, Tulsinol E, Tulsinol F, and Tulsinol G
Coumarins	Ocimarin, aesculetin, and aesculin
Terpenoids	Sesquiterpenoids (β-caryophyllene and 4,5-epoxy-caryophyllene), abietane diterpenoid (carnosic acid), oleane triterpenoids (oleanolic acid, β-Amyrin-glucopyranoside) and ursane triterpenoids (ursolic acid, urs-12-en-3β,6β,20β-triol-28-oic acid), 16-hydroxy-4,4,10,13-tetramethyl-17-(4-methylpentyl)tetradecahydro-1*H*-cyclopenta[a]phenanthren-3(2*H*)-one, tricyclic sesquiterpenoid-2-(hydroxymethyl)-5,5,9-trimethylcyclo[7.2.0.0] undecan-2-ol, β-caryophyllene, elemene, α-humulene or α-caryophyllene, germacrene-A, trans-α-bergamotene, and 5β-hydroxycaryophyllene
Steroids	β-sitosterol, β-sitosterol-3-O-β-D-glucopyranoside, stigmasterol, and campesterol

TABLE 6.1   *(Continued)*

Nature of Chemical Constituents	Constituents Reported
Essential oils	Eugenol or methyl eugenol, and methyl chavicol
Fixed oil (non-volatile oil)	Linoleic acid, α-linolenic acid, oleic acid, palmitic acid, and stearic acid
Fatty acid derivatives	Palmityl glucoside and sanctumoic acid
Polysaccharides	Rhamnose, xylose, arabinose, glucose, and galactose

6.6   BIOLOGICAL ACTIVITIES OF *OCIMUM SANCTUM*

6.6.1   ANTICANCER ACTIVITY

O. sanctum has been found to possess anticancer activity. Dichloro-methane and hydromethanolic extracts of aerial parts of *O. sanctum* found to possess following compounds like terpenoids (ursolic acid, oleanolic acid, betulinic acid, stigmasterol, and, β-caryophyllene oxide), lignans [(-)-rabdosiin and shimobashiric acid C], flavonoids (luteolin and 7-O-β-D-glucuronide, apigenin 7-O-β-D-glucuronide), phenolics [(*E*)-*p*-coumaroyl 4-O-β-D-glucoside, 3-(3,4dihydroxyphenyl) lactic acid, protocatechuic acid, and vanillic acid]. The compounds were evaluated for their cytotoxicity against MCF-7 (breast cancer), SKBR3 (breast cancer), HCT-116 (colon cancer), and normal peripheral blood mononuclear cells (PBMCs). (-)-rabdosiin was the most cytotoxic compound found against cancer cell lines considered. It exerted marginal cytotoxicity against PBMCs. Doxorubicin was used as the positive control. The IC_{50} values were 75 ± 2.12 µg/mL for MCF-7, 83 ± 3.54 µg/mL for SKBR3 (breast cancer), and 84 ± 7.78 µg/mL for HCT-116 [14] (Figure 6.1).

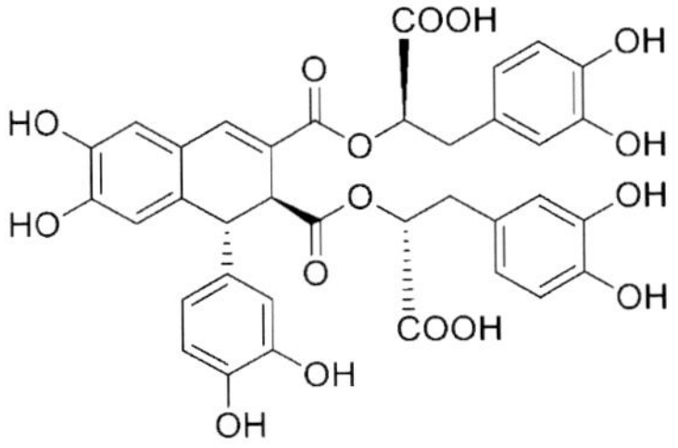

(-)-Rabdosiin

FIGURE 6.1   Lignan present in *Ocimum sanctum*.

A flavonoid named vicenin-2 which is an active constituent of Tulsi was evaluated for anticancer potential and was found effective against carcinoma of the prostate (CaP) [15]. Orientin, along with luteolin were also evaluated for the same. Vivenin-2 was found to be more potent than orientin and luteolin. The prostate cancer cell lines used in the study were PC-3, DU-145 (androgen-independent), and LNCaP (androgen-dependent). Vicenin-2 exhibited anti-proliferative, anti-angiogenic, and pro-apoptotic effects in all cell lines. The effects were independent of their androgen sensitivity or p53 status. Androgens have a critical role in the growth and maintenance of normal and cancer tissue. Therefore, androgen-dependent cancer cells (CRC) undergo apoptosis during androgen-ablation therapy. The therapy mainly focuses on reducing endogenous androgen levels which can be achieved by directly blocking androgen receptor activity. However, androgen-independent tumors or CRC perpetually escape these treatments with highly incomprehensible mechanisms like androgen receptor amplification leading to mutations, overexpression of the receptors and, de-regulated expression of androgen receptor co-regulators [16–19]. Vicenin-2 was found to inhibit EGFR/Akt/mTOR/ p70S6K pathway *in vitro*. Prostate cancer progression and metastasis are associated with EGFR and PI3K/Akt/mTOR activation. EGFR (170 kDa) is the epidermal growth factor receptor and its overexpression is observed in various malignancies including prostate cancer. It is a proto-oncogene and transmembrane receptor. EGFR, when interacts with ligand, triggers a series of intracellular signaling pathways and as a consequence DNA synthesis, cell proliferation takes place [20]. Thereby, attenuation of EGFR by vicenin 2 affirms the effectiveness against CaP. In the same way, PI3K/AKT/mTOR pathway inhibition by vicenin-2 also shows potential against prostate cancer. This is an intracellular signaling pathway occurs in the cell which is significant in regulating the cell cycle. Cellular quiescence, proliferation, cancer, and longevity are associated with this pathway [21, 22]. Vicenin-2 also found to down-regulate c-Myc, cyclin D1, cyclin B1, CDK4, PCNA, and hTERT *in vitro*. c-Myc is a regulator gene and encoded a protein of this gene plays a role in cell cycle progression, apoptosis, and cellular transformation. Overexpression of c-Myc relates to cellular proliferation. The down-regulation of c-Myc by vicenin-2 shows its anti-proliferative effect. Cyclin D1 and cyclin B1 key cell-cycle regulatory protein. Cyclin D1 is essential in the G1 to S transition of the cell cycle. It binds to CDK4 (cyclin-dependent kinase 4), which is a cell division protein kinase. Binding of cyclin D1 to CDK4 results in phosphorylation and inactivates retinoblastoma protein pRb, which is a tumor suppressor protein. Therefore, the down-regulation of aberrant cyclin D1 expression prevents excessive cell growth by hindering cell cycle progression. Cyclin B1 binds to CDK1 (cyclin-dependent kinase 1) and essential for G2/M transition [23–26]. Loss of proliferating cell nuclear antigen (PCNA) may contribute in growth arrest and human telomerase reverse transcriptase (hTERT) expression has been observed in various types of human cancer, so decreased levels of hTERT by vicenin-2 is another indication of its anticancer effects [27, 28]. Vicenin-2, while treated with docetaxel (current drug of choice in androgen-independent CaP) produces synergistic inhibition of the growth of prostate cancer *in vivo* (mice xenografts). Vicenin-2 can be orally administered because the serum level of 2.6 ± 0.3 micromol/L after oral administration in mice was observed in the same study. IC_{50} values of vicenin-2, luteolin, and orientin against PC3 cell line were found to be 26 ± 3 µM, 48 ± 3 µM, and 98 ± 7 µM, respectively. IC_{50} values of the same against the DU145 cell line were 25 ± 3 µM, 57 ± 5 µM, and 107 ± 6 µM, respectively. Against LNCap cell, line the IC_{50} values were 44 ± 3 µM, 78 ± 6 µM, and 124 ± 7 µM, respectively for vicenin-2, luteolin, and orientin. 2-(hydroxymethyl)-5,5,9-trimethyltricyclo[7.2.0.0] undecan-2-ol, a tricyclic sesquiterpenoid was isolated from the leaves of Tulsi. It was evaluated against MCF-7 (breast cancer) cell line. IC_{50} value was 30 ± 0.5 µM [29]. Quercetin of *O. sanctum* in conjugation with polyethylene glycol coated nickel nanoparticles (NPs) has been found to possess anticancer activity human breast cancer MCF-7 cells. It exerted a dose-dependent anticancer effect against MCF-7 cells. The IC_{50} value was found to be 6.25 µg/ml. It produced apoptosis in MCF-7 cells. The IC_{50} value of quercetin alone was 50 µg/ml [30].

6.6.2 ANTIOXIDANT ACTIVITY

Reactive oxygen species (ROS) or oxygen-free radical formation takes place in aerobic cells due to oxidative stress. However, this is countered by complex systems of endogenous antioxidants (AOs) such as glutathione and enzymes like catalase (CAT) and superoxide dismutase (SOD). It is also inhibited by the dietary AOs vitamin C, and vitamin E. If oxidative stress remains uncounteract, the free radicals generated thereby may interact with lipids to produce hydroperoxides and peroxides (Figure 6.2).

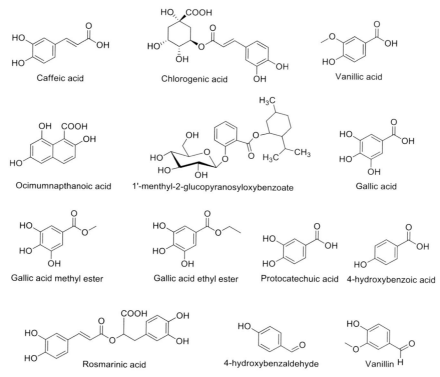

FIGURE 6.2   Phenolic compounds present in *Ocimum sanctum*.

This may lead to harmful consequences in the biological system resulting in cancer. Therefore, phytoconstituents with free-radical scavenging capability can have a vital impact on oxidative stress and may prevent the aftermath of oxidative stress [31–33]. Flavonoids have been found to effectively contravene the free radical formation. Extraction of the fresh leaves and stems of *O. sanctum* and evaluation of antioxidant activity were done in a study. Subsequent purification of the extract (antioxidant bioassay-directed) yielded that Tulsi contains flavonoids like cirsilineol, cirsimaritin, isothymusin, and isothymonin which were found to have moderate to strong antioxidant activity at 10 µM concentration, while tested using the liposome oxidation model. The activity was comparable to the standard used like TBHQ (tert-butylhydroquinone) and BHT (butyl-ated hydroxytoluene). Rosmarinic acid, which is phenolic acid found in holy basil in the same study, demonstrated better antioxidant activity (at 10 μM) than the standard vitamin E used as a positive control in the liposome oxidation model. Essential oil (EO), eugenol (at 10 μM) too possess better antioxidant activity than the standard TBHQ and BHT in the same model used in the study [6]. Administration of ethanolic extract of *O. sanctum* could attenuate the effects of noise-induced oxidative stress/damage to the brain region in Wistar strain male albino rats. The attenuating effect may be due to the presence of a large quantity of phytochemicals such as phenolics and flavanoids in *O. sanctum* [34] (Figure 6.3).

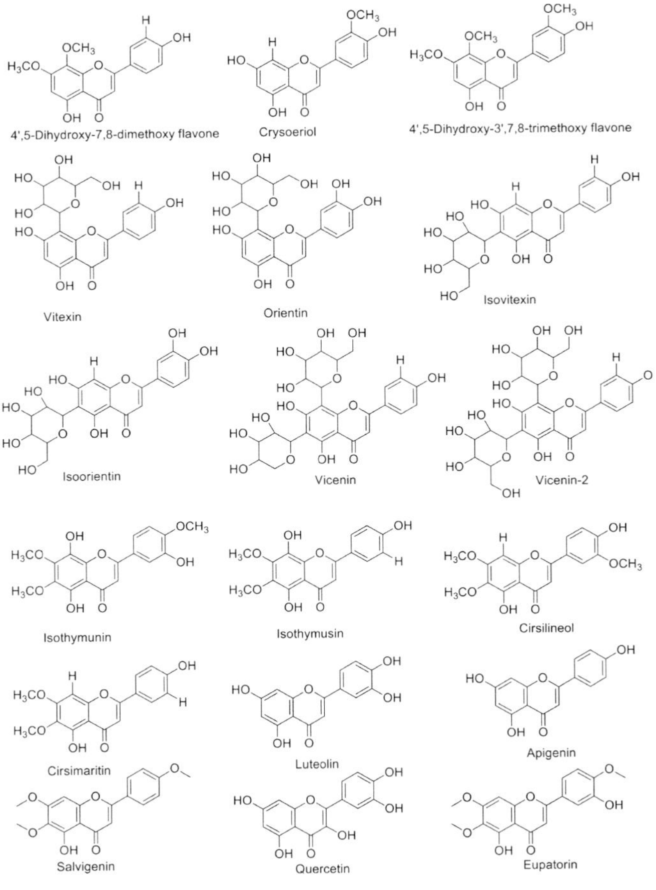

FIGURE 6.3 Flavonoids and glycosides present in *Ocimum sanctum*.

6.6.3   LEISHMANICIDAL ACTIVITY

Leishmaniasis is a protozoan parasitic disease caused by Leishmania parasites (*Leishmania donovani, Leishmania tropica*) [35]. The extracts of dried leaves of Tulsi exerted leishmanicidal activity. The activity was evaluated against promastigotes of *Leishmania major*. Ethyl acetate fraction and n-butanol fraction showed positive results. The activity is due to the presence of phenylpropanoids (ferulaldehyde, bieugenol, and dehydrodieugenol B), flavonoids (luteolin, apigenin, 4',5-dihydroxy-7,8-dimethoxy flavone, and 4',5-dihydroxy-3',7,8-trimethoxy flavone), neolignans, and triterpenoids like ursolic acid and oleanolic acid. Ferulaldehyde (IC_{50} 0.9 µg/ml) and ursolic acid (IC_{50} 2.2 µg/ml) showed strong leishmanicidal activity. IC_{50} values for other compounds ranging from 9 µg/ml to 90 µg/ml [36, 37] (Figures 6.4 and 6.5).

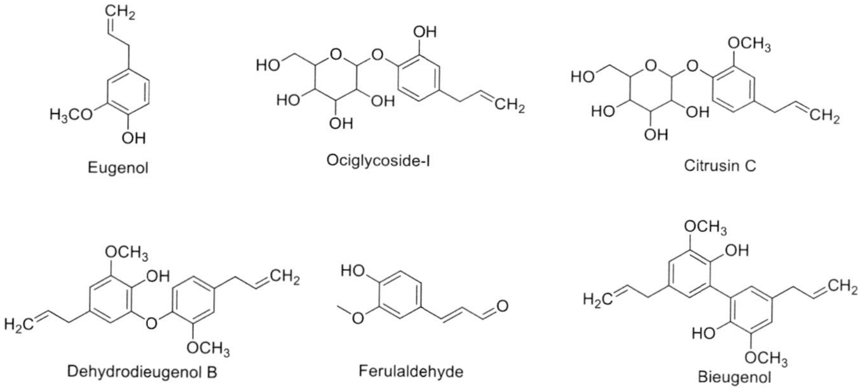

FIGURE 6.4   Phenylpropanoids present in *Ocimum sanctum*.

6.6.4   ANTIBACTERIAL ACTIVITY

Ocimum flavonoids; orientin and vicenin were isolated from the aqueous extract of fresh leaves of Tulsi. Both flavonoids demonstrated antibacterial activity against gram-positive bacteria (*Staphylococcus aureus* and *Staphylococcus cohnii*) and gram-negative bacteria (*Escherichia coli, Proteus*, and *Klebsialla pneumonia*). Zones of inhibition (ZOI) of 20.12, 20.75, 20.95, 19.55, and 20.1 mm at the concentration of 400 mg/ml were found against *Escherichia coli, Proteus, Staphylococcus aureus, Staphylococcus cohnii*, and *Klebsialla pneumonia* [38] (Figures 6.6 and 6.7).

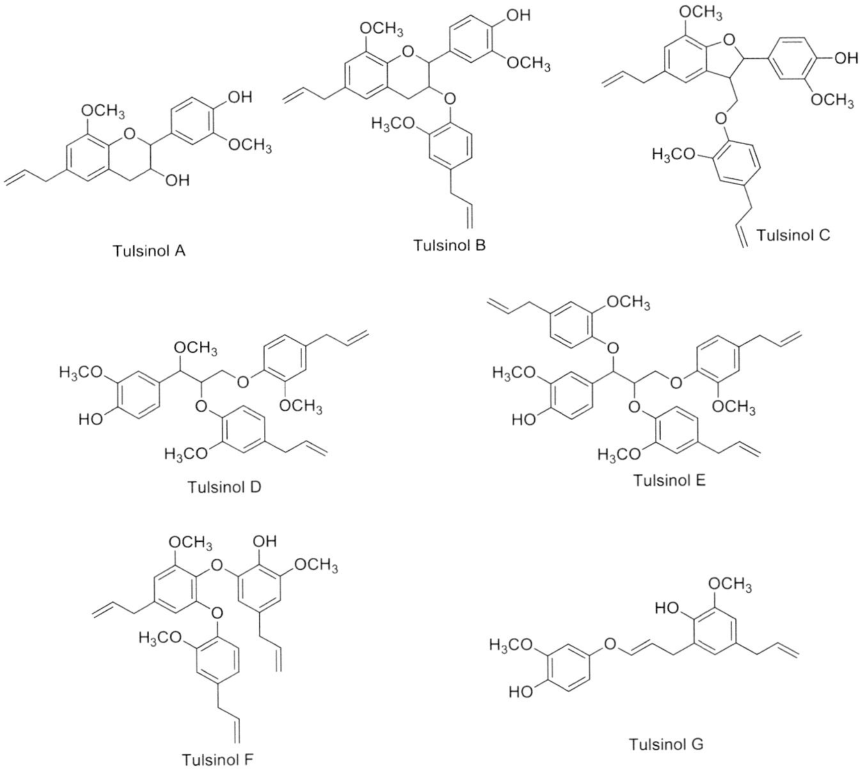

FIGURE 6.5 Neolignans present in *Ocimum sanctum*.

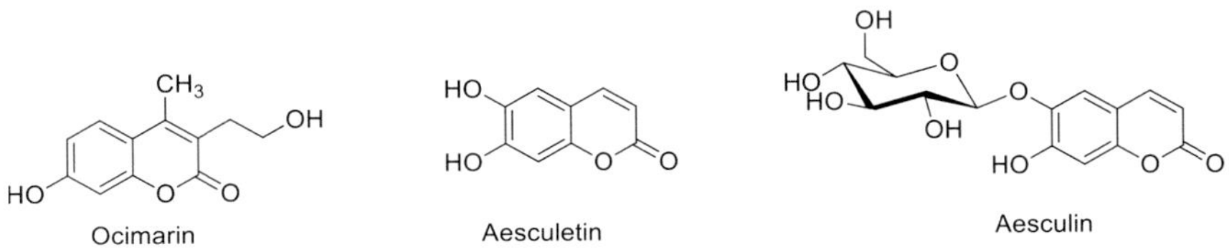

FIGURE 6.6 Coumarins present in *Ocimum sanctum*.

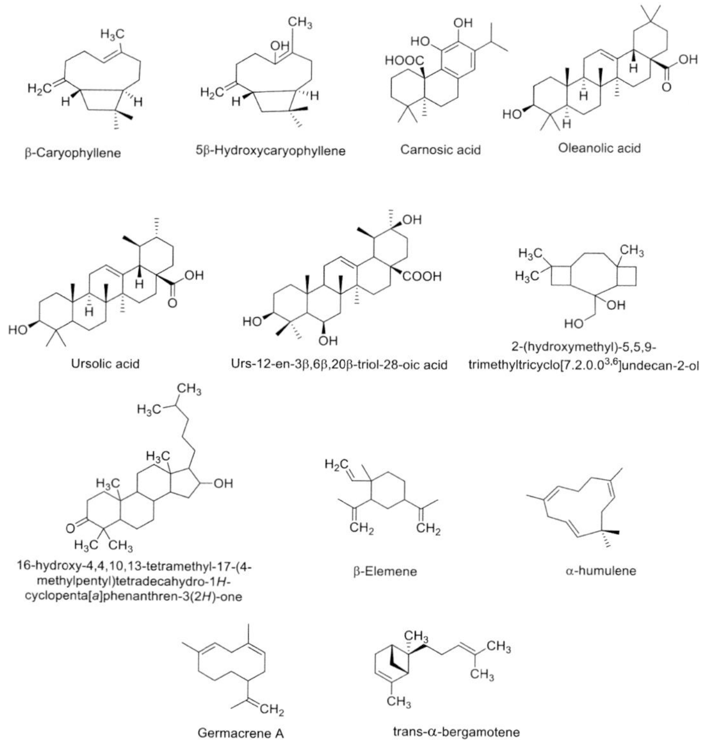

β-Caryophyllene   5β-Hydroxycaryophyllene   Carnosic acid   Oleanolic acid

Ursolic acid   Urs-12-en-3β,6β,20β-triol-28-oic acid   2-(hydroxymethyl)-5,5,9-trimethyltricyclo[7.2.0.0³,⁶]undecan-2-ol

16-hydroxy-4,4,10,13-tetramethyl-17-(4-methylpentyl)tetradecahydro-1H-cyclopenta[a]phenanthren-3(2H)-one   β-Elemene   α-humulene

Germacrene A   trans-α-bergamotene

FIGURE 6.7   Terpenoids present in *Ocimum sanctum*.

6.6.5   ANTIFUNGAL ACTIVITY

Antifungal activity of the EO of *Ocimum sp.* has been evaluated against fluconazole-sensitive and fluconazole-resistant *Candida* isolates. The fungal strains incorporated were fluconazole sensitive *Candida* species (*C. albicans, C. tropicalis,* and *C. parapsilosis*) and fluconazole-resistant *C. albicans, C. tropicalis* isolates; and *C. krusei* and *C. glabrata* which are intrinsically resistant to fluconazole. The MIC values were found to be in the range from 0.1 µl/ml to 0.5 µl/ml. The EO also responded to synergistic effects with fluconazole and ketoconazole. The GC-MS analysis confirmed the presence of D-limonene, linalool, α-farnesene, caryophyllene, menthol, methyl chavicol, α-citral, carvone, *p*-methoxy cinnamaldehyde, and α-caryophyllene [39] (Figures 6.8–6.10).

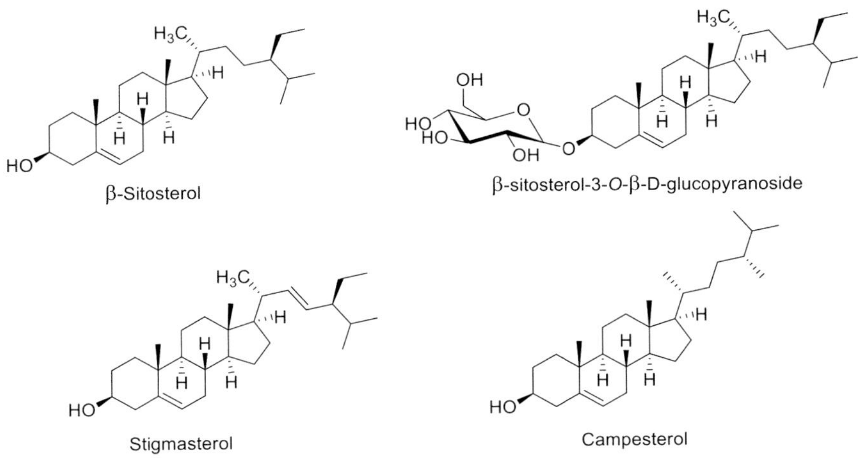

FIGURE 6.8 Steroids of *Ocimum sanctum*.

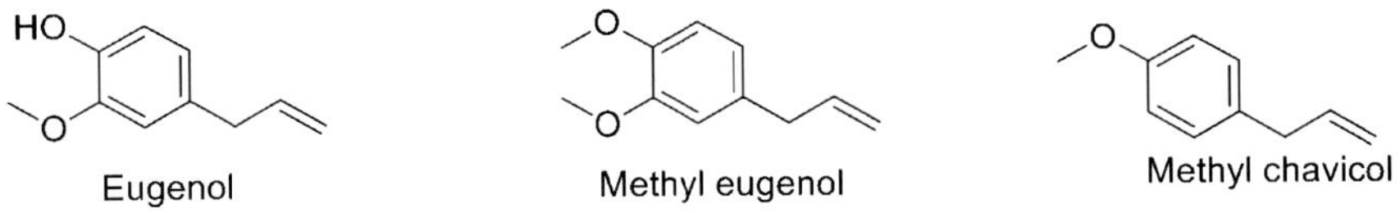

FIGURE 6.9 Essential oils of *Ocimum sanctum*.

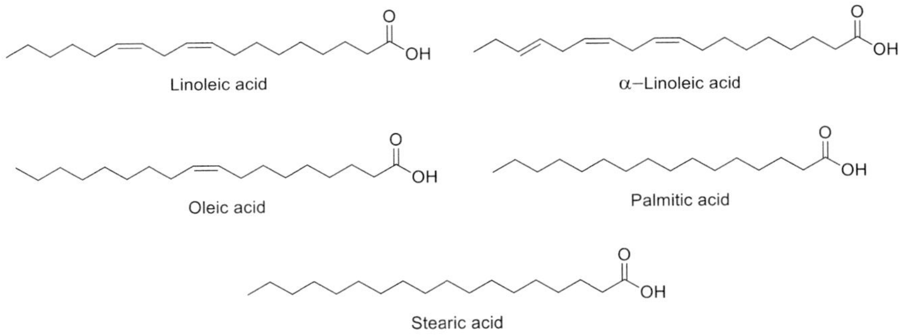

FIGURE 6.10 Saturated and unsaturated fatty acids of *Ocimum sanctum*.

The leaf extracts of *Ocimum sp.* were found to be effective against clinically isolated dermatophytic fungi (*Trichophyton mentagrophytes, Trichophyton rubrum, Microsporum gypseum, Microsporum nanum*, and *Epidermophyton flocossum*). MIC (minimum inhibitory concentration) values for aqueous extract were found to be ranging from 200 to 416 µg/

ml. For alcoholic extract, the MIC values were in the range of 166 to 383 μg/ml against the selected fungal species [40]. Methanolic extract of *O. sanctum* was also found to be effective against *C. albicans* strains isolated from the high vaginal swab, urine, and catheter tip. The extract also demonstrated synergistic activity with fluconazole against the selected strains [41].

6.6.6   ANTI-INFLAMMATORY ACTIVITY

The ethanolic extract of *Ocimum sp.* has been found to possess significant suppression capability of the NF-κB expression (79.3 ± 9.6%). It also inhibited IL-6 secretion (54.7 ± 3.1%) [42]. The transcription factor NF-κB is associated with innate and adaptive immune functions and functions as a critical mediator of inflammatory responses [43, 44]. Interleukin (IL)-6 is also related to inflammation and produced at the site of inflammation. IL-6 is responsible for the stimulatory effects on T-cells and B-cells, thus supporting chronic inflammatory responses [45, 46]. Flavonoids (apigenin, isothymusin, isothymonin, cirsimaritin, cirsilineol, and 4′,5-dihydroxy-3′,7,8-trimethoxy flavone), phenolic acid (rosmarinic acid), and phenylpropanoid (eugenol) of Tulsi demonstrated anti-inflammatory activity through cyclooxygenase (COX) inhibition. Among them, eugenol demonstrated 97% cyclooxygenase-1 (COX-1) inhibitory activity at 1 μM concentration. Other compounds displayed moderate COX-1 inhibitory activity. Interestingly, isothymusin was found inactive against COX-1 [6].

6.6.7   ANTI-DIABETIC ACTIVITY

A tetracyclic triterpenoid anti-diabetic component of Tulsi named 16-hydroxy-4,4,10,13-tetramethyl-17-(4-methylpentyl)-hexadeca-hydrocyclopenta[a]-phenanthren-3-one was identified by the bioactivity guided fractionation of hydroalcoholic extract of the aerial part. The anti-diabetic activity was evaluated in alloxan-induced diabetic rats by estimating the serum glucose level and lipid parameters. The compound has been found to significantly decrease serum glucose level, total cholesterol (TC), triglycerides (TG), LDL cholesterol, and increase in serum HDL cholesterol level in treated diabetic rats [47].

6.6.8   ANTI-ULCEROGENIC AND ULCER-HEALING PROPERTIES

The anti-ulcerogenic activity of extract of leaves of *O. sanctum* was found at a dose of 100 mg/kg. The animal models incorporated in the study were cold restraint, aspirin, alcohol, and pyloric ligation induced gastric ulcer models in Sprague-Dawley rats. Guinea pigs were used for histamine-induced duodenal ulcer and ulcer-healing activity was done on acetic acid-induced chronic ulcer model. The extract decreased the incidence of ulcers and also enhanced the healing of ulcers. It significantly reduced free, total acidity and peptic activity by and increased mucin secretion. It was also found to completely heal the ulcers within 20 days of treatment in acetic acid-induced chronic ulcer model. The anti-ulcer effect of the extract may be due to its cytoprotective effect rather than anti-secretory activity [48].

6.6.9   ADAPTOGENIC ACTIVITY/ANTISTRESS ACTIVITY

Ocimumosides (A and B) and ocimarin were isolated from an extract of the leaves of holy basil (*O. sanctum*). They were found to possess anti-stress activity at a dose of 40 mg/kg body weight. The activity was evaluated against acute stress-induced biochemical changes in male Sprague-Dawley rats. These *Ocimum* components normalized hyperglycemia, plasma corticosterone, plasma creatine kinase, and adrenal hypertrophy, which are considered as the parameter for stress [49] (Figures 6.11a and 6.11b).

6.6.10   MISCELLANEOUS BIOLOGICAL ACTIVITIES

Ocimum sanctum was found to possess prebiotic activity, with abundant growth of *Lactobacillus* and *Bifidobacterium*. It can be comparable with the well-known prebiotic, fructo-oligosaccharide (FOS) [50]. Ethanolic extract of *Ocimum sp.* was found to enhance cognitive ability in the rat *in vivo* model. The administration of 100 mg/kg of dose induced the expression of choline acetyltransferase in rat brain [51]. The decline in cognitive function is correlated with the decreased choline acetyltransferase activity and loss of cholinergic neurons. As a consequence, the damage to the central cholinergic system and the reduction in spatial memory ability take place [52, 53]. The increased choline acetyltransferase expression by an ethanolic extract of *Ocimum sp.* may enhance the cognitive ability in rats.

The cognitive ability may be simulative to humans too. The extracts of
Ocimum sp. were found to inhibit collagenase, elastase, and hyaluronidase
activities *in vitro* demonstrating its skin anti-aging potential. Ethanolic
extract demonstrated MMP-1 (collagenase) and hyaluronidase inhibitory
activities with the values of $77.7 \pm 9.0\%$ and $79.3 \pm 9.6\%$, respectively.

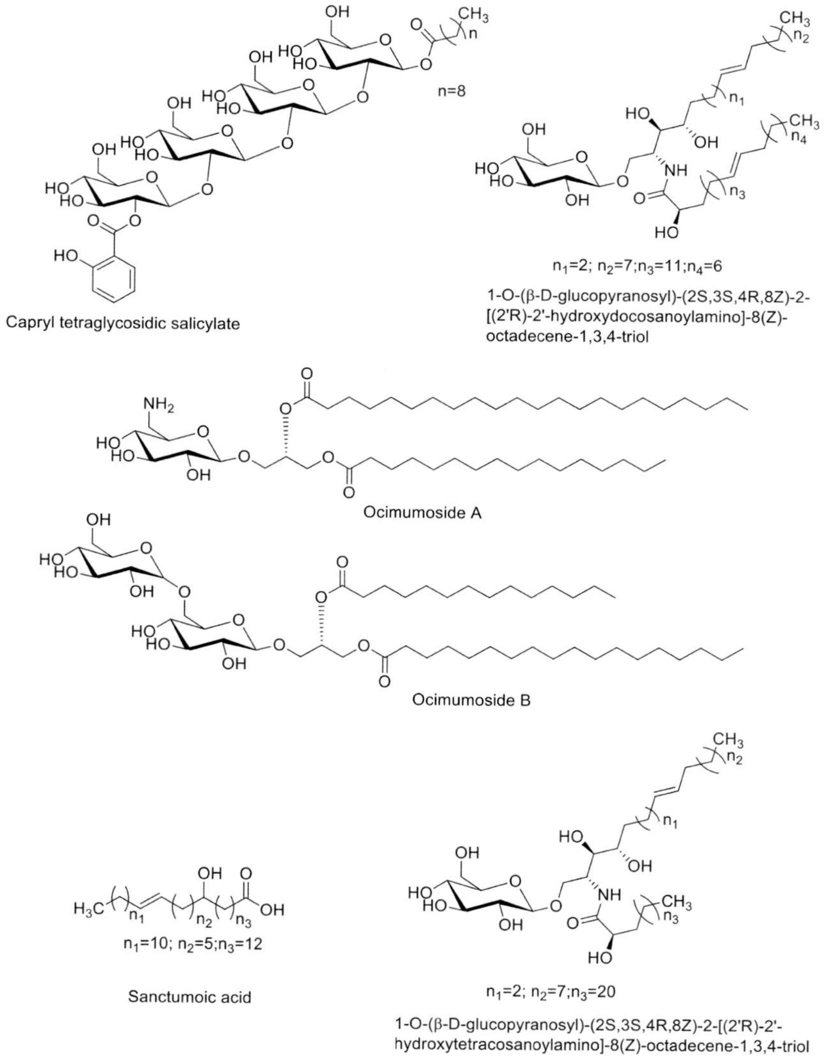

Capryl tetraglycosidic salicylate

$n_1=2; n_2=7; n_3=11; n_4=6$

1-O-(β-D-glucopyranosyl)-(2S,3S,4R,8Z)-2-
[(2'R)-2'-hydroxydocosanoylamino]-8(Z)-
octadecene-1,3,4-triol

Ocimumoside A

Ocimumoside B

$n_1=10; n_2=5; n_3=12$

Sanctumoic acid

$n_1=2; n_2=7; n_3=20$

1-O-(β-D-glucopyranosyl)-(2S,3S,4R,8Z)-2-[(2'R)-2'-
hydroxytetracosanoylamino]-8(Z)-octadecene-1,3,4-triol

FIGURE 6.11   (a–b) Fatty acid derivatives of *Ocimum sanctum.*

The loss/breakdown of collagen is associated with aging skin. Thus, inhibition of collagenase may retard skin aging. Rosmarinic acid may be attributed to this skin anti-aging potential [42]. Nano-encapsulated *Ocimum sp.* EO was found to be effective against fungi contaminating herbal raw materials and aflatoxin B1 (AFB1) secretion. Chitosan was used for encapsulation. Therefore, it has anti-fungal and anti-aflatoxigenic potential [54]. *O. sanctum* could also be used for the biogenic synthesis of silver nanoparticles (AgNPs). The optimized AgNPs into the carbopol gel base were found to possess wound healing activity in the rat model of skin wound healing. The antibacterial inhibitory efficiency of this nanogel was comparable to the commercial product, which is effective against the *Staphylococcus aureus, E. coli, and Pseudomonas aeruginosa* [55].

6.7   CONCLUSION

It can be unquestionably granted that *O. sanctum* is a significant medicinal plant with a diverse range of phytoconstituents. Marketed ayurvedic formulations have *Ocimum sp.* as a major ingredient for the treatment of a variety of illnesses. Tulsi in the form of tea and their decoctions are already popular in India. With the growing demand for herbal dietary supplement and herbal medicines, it can be surely predicted that *Ocimum sp.* is an herb of great industrial significance. However, in the near future, the active constituents of Tulsi individually and/or in combination may be useful in the treatment of various diseases.

KEYWORDS

- Ayurveda
- carcinoma of the prostate
- interleukin
- minimum inhibitory concentration
- *Ocimum sanctum*
- tert-butyl hydroquinone
- zones of inhibition

REFERENCES

1. Rossiter, S. E., Fletcher, M. H., & Wuest, W. M., (2017). Natural products as platforms to overcome antibiotic resistance. *Chemical Reviews, 117*(19), 12415–12474.
2. Manna, K., Debnath, B., Das, M., & Marwein, S., (2016). A comprehensive review on pharmacognostical investigation and pharmacology of *Typhonium trilobatum. The Natural Products Journal, 6*(3), 172–178.
3. Davison, E. K., & Sperry, J., (2017). Natural products with heteroatom-rich ring systems. *Journal of Natural Products, 80*(11), 3060–3079.
4. Debnath, B., Singh, W. S., Das, M., Goswami, S., Singh, M. K., Maiti, D., & Manna, K., (2018). Role of plant alkaloids on human health: A review of biological activities. *Materials Today Chemistry, 9*, 56–72.
5. Cordell, G. A., & Colvard, M. D., (2012). Natural products and traditional medicine: Turning on a paradigm. *Journal of Natural Products, 75*(3), 514–525.
6. Kelm, M. A., Nair, M. G., Strasburg, G. M., & DeWitt, D. L., (2000). Antioxidant and cyclooxygenase inhibitory phenolic compounds from *Ocimum sanctum* Linn. *Phytomedicine, 7*(1), 7–13.
7. Prakash, P., & Gupta, N., (2005). Therapeutic uses of *Ocimum sanctum* Linn. (Tulsi) with a note on eugenol and its pharmacological actions: A short review. *Indian Journal of Physiology and Pharmacology, 49*(2), 125.
8. Vats, V., Grover, J. K., & Rathi, S. S., (2002). Evaluation of anti-hyperglycemic and hypoglycemic effect of *Trigonella foenum-graecum* Linn., *Ocimum sanctum* Linn., and *Pterocarpus marsupium* Linn. in normal and alloxanized diabetic rats. *Journal of Ethnopharmacology, 79*(1), 95–100.
9. Pattanayak, P., Behera, P., Das, D., & Panda, S. K., (2010). *Ocimum sanctum* Linn. A reservoir plant for therapeutic applications: An overview. *Pharmacognosy Reviews, 4*(7), 95.
10. Shetty, S., Udupa, S., & Udupa, L., (2008). Evaluation of antioxidant and wound healing effects of alcoholic and aqueous extract of *Ocimum sanctum* Linn. in rats. *Evidence-Based Complementary and Alternative Medicine, 5*(1), 95–101.
11. Singh, D., & Chaudhuri, P. K., (2018). A review on phytochemical and pharmacological properties of Holy basil (*Ocimum sanctum* L.). *Industrial Crops and Products, 118*, 367–382.
12. Kavyashree, M. R., Harini, A., Hegde, P. L., & Pradeep, (2019). A review on Tulasi (*Ocimum sanctum* Linn.). *Journal of Drug Delivery and Therapeutics, 9*(2-s), 562–569.
13. The Ayurvedic Pharmacopoeia of India, (2016). *Pharmacopoeia Commission for Indian Medicine and Homoeopathy* (Part-I, 1st edn., Vol. IX, pp. 99–105). Ghaziabad, Ministry of AYUSH, Government of India.
14. Flegkas, A., Milosević, I. T., Barda, C., Samara, P., Tsitsilonis, O., & Skaltsa, H., (2019). Anti-proliferative activity of (-)-Rabdosiin isolated from *Ocimum sanctum* L. *Medicines, 6*(1), 37.
15. Nagaprashantha, L. D., Vatsyayan, R., Singhal, J., Fast, S., Roby, R., Awasthi, S., & Singhal, S. S., (2011). Anti-cancer effects of novel flavonoid vicenin-2 as a single

agent and in synergistic combination with docetaxel in prostate cancer. *Biochemical Pharmacology, 82*(9), 1100–1109.

16. Watson, P. A., Arora, V. K., & Sawyers, C. L., (2015). Emerging mechanisms of resistance to androgen receptor inhibitors in prostate cancer. *Nature Reviews Cancer, 15*(12), 701–711.

17. Gundem, G., Van, L. P., Kremeyer, B., Alexandrov, L. B., Tubio, J. M., Papaemmanuil, E., et al., (2015). The evolutionary history of lethal metastatic prostate cancer. *Nature, 520*(7547), 353–357.

18. Tan, M. E., Li, J., Xu, H. E., Melcher, K., & Yong, E. L., (2015). Androgen receptor: Structure, role in prostate cancer and drug discovery. *Acta Pharmacologica Sinica, 36*(1), 3–23.

19. Steinestel, J., Luedeke, M., Arndt, A., Schnoeller, T. J., Lennerz, J. K., Wurm, C., et al., (2019). Detecting predictive androgen receptor modifications in circulating prostate cancer cells. *Oncotarget, 10*(41), 4213–4223.

20. Day, K. C., Hiles, G. L., Kozminsky, M., Dawsey, S. J., Paul, A., Broses, L. J., Shah, R., Kunja, L. P., Hall, C., Palanisamy, N., & Daignault-Newton, S., (2017). HER2 and EGFR over expression support metastatic progression of prostate cancer to bone. *Cancer Research, 77*(1), 74–85.

21. Chang, L., Graham, P. H., Ni, J., Hao, J., Bucci, J., Cozzi, P. J., & Li, Y., (2015). Targeting PI3K/Akt/mTOR signaling pathway in the treatment of prostate cancer radio resistance. *Critical Reviews in Oncology/Hematology, 96*(3), 507–517.

22. Lim, H. J., Crowe, P., & Yang, J. L., (2015). Current clinical regulation of PI3K/PTEN/Akt/mTOR signaling in treatment of human cancer. *Journal of Cancer Research and Clinical Oncology, 141*(4), 671–689.

23. Hydbring, P., Malumbres, M., & Sicinski, P., (2016). Non-canonical functions of cell cycle cyclins and cyclin-dependent kinases. *Nature Reviews Molecular Cell Biology, 17*(5), 280–292.

24. Zhao, B., & Burgess, K., (2019). PROTACs suppression of CDK4/6, crucial kinases for cell cycle regulation in cancer. *Chemical Communications, 55*(18), 2704–2707.

25. Das, M., & Manna, K., (2016). Chalcone scaffold in anticancer armamentarium: A molecular insight. *Journal of Toxicology*. http://dx.doi.org/10.1155/2016/7651047 (accessed on 14 May 2020).

26. Żuryń, A., Krajewski, A., Klimaszewska-Wiśniewska, A., Grzanka, A., & Grzanka, D., (2019). Expression of cyclin B1, D1, and K in non-small cell lung cancer H1299 cells following treatment with sulforaphane. *Oncology Reports, 41*(2), 1313–1323.

27. Jurikova, M., Danihel, L'., Polák, Š., & Varga, I., (2016). Ki67, PCNA, and MCM proteins: Markers of proliferation in the diagnosis of breast cancer. *Acta Histochemica, 118*(5), 544–552.

28. Stögbauer, L., Stummer, W., Senner, V., & Brokinkel, B., (2019). Telomerase activity, TERT expression, hTERT promoter alterations, and alternative lengthening of the telomeres (ALT) in meningiomas: A systematic review. *Neurosurgical Review*, pp. 1–8.

29. Singh, D., Chaudhuri, P. K., & Darokar, M. P., (2014). New antiproliferative tricyclic sesquiterpenoid from the leaves of *Ocimum sanctum*. *Helv. Chim. Acta, 97*, 708–711.

30. Rameshthangam, P., & Chitra, J. P., (2018). Synergistic anticancer effect of green synthesized nickel nano particles and quercetin extracted from *Ocimum sanctum* leaf extract. *Journal of Materials Science and Technology, 34*(3), 508–522.

31. Shahidi, F., & Ambigaipalan, P., (2015). Phenolics and polyphenolics in foods, beverages, and spices: Antioxidant activity and health effects: A review. *Journal of Functional Foods, 18*, 820–897.

32. Zou, Z., Xi, W., Hu, Y., Nie, C., & Zhou, Z., (2016). Antioxidant activity of citrus fruits. *Food Chemistry, 196*, 885–896.

33. Nile, S. H., Nile, A. S., & Keum, Y. S., (2017). Total phenolics, antioxidant, antitumor, and enzyme inhibitory activity of Indian medicinal and aromatic plants extracted with different extraction methods. *Three Biotech., 7*(1), 76.

34. Samson, J., Sheeladevi, R., & Ravindran, R., (2007). Oxidative stress in brain and antioxidant activity of *Ocimum sanctum* in noise exposure. *Neurotoxicology, 28*(3), 679–685.

35. Taxy, J. B., Goldin, H. M., Dickie, S., & Cibull, T., (2019). *Cutaneous leishmaniasis*. The *American Journal of Surgical Pathology, 43*(2), 195–200.

36. Suzuki, A., Shirota, O., Mori, K., Sekita, S., Fuchino, H., Takano, A., & Kuroyanagi, M., (2009). Leishmanicidal active constituents from Nepalese medicinal plant Tulsi (*Ocimum sanctum* L.). *Chem. Pharm. Bull., 57*, 245–251.

37. Mahajan, N., Rawal, S., Verma, M., Poddar, M., & Alok, S., (2013). A phytopharmacological overview on *Ocimum* species with special emphasis on *Ocimum sanctum*. *Biomedicine and Preventive Nutrition, 3*(2), 185–192.

38. Ali, H., & Dixit, S., (2012). *In vitro* antimicrobial activity of flavanoids of *Ocimum sanctum* with synergistic effect of their combined form. *Asian Pacific Journal of Tropical Disease, 2*, S396–S398.

39. Amber, K., Aijaz, A., Immaculata, X., Luqman, K. A., & Nikhat, M., (2010). Anticandidal effect of *Ocimum sanctum* essential oil and its synergy with fluconazole and ketoconazole. *Phytomedicine, 17*(12), 921–925.

40. Balakumar, S., Rajan, S., Thirunalasundari, T., & Jeeva, S., (2011). Antifungal activity of *Ocimum sanctum* Linn. (Lamiaceae) on clinically isolated dermatophytic fungi. *Asian Pacific Journal of Tropical Medicine, 4*(8), 654–657.

41. Zaidi, K. U., Shah, F., Parmar, R., & Thawani, V., (2018). Anticandidal synergistic activity of *Ocimum sanctum* and fluconazole of azole resistance strains of clinical isolates. *Journal De Mycologie Medicale, 28*(2), 289–293.

42. Chaiyana, W., Anuchapreeda, S., Punyoyai, C., Neimkhum, W., Lee, K. H., Lin, W. C., Lue, S. C., Viernstein, H., & Mueller, M., (2019). *Ocimum sanctum* Linn. as a natural source of skin anti-aging compounds. *Industrial Crops and Products, 127*, 217–224.

43. Liu, T., Zhang, L., Joo, D., & Sun, S. C., (2017). NF-κB signaling in inflammation. *Signal Transduction and Targeted Therapy, 2*, 17023.

44. Taniguchi, K., & Karin, M., (2018). NF-κB, inflammation, immunity, and cancer: Coming of age. *Nature Reviews Immunology, 18*(5), 309.

45. Gabay, C., (2006). Interleukin-6 and chronic inflammation. *Arthritis Research and Therapy, 8*(2), S3.

46. Jones, S. A., & Jenkins, B. J., (2018). Recent insights into targeting the IL-6 cytokine family in inflammatory diseases and cancer. *Nature Reviews Immunology*, p. 1.

47. Patil, R., Patil, R., Ahirwar, B., & Ahirwar, D., (2011). Isolation and characterization of anti-diabetic component (bioactivity-guided fractionation) from *Ocimum sanctum* L. (Lamiaceae) aerial part. *Asian Pacific Journal of Tropical Medicine, 4*(4), 278–282.

48. Dharmani, P., Kuchibhotla, V. K., Maurya, R., Srivastava, S., Sharma, S., & Palit, G., (2004). Evaluation of anti-ulcerogenic and ulcer-healing properties of *Ocimum sanctum* Linn. *Journal of Ethno Pharmacology, 93*(2–3), 197–206.

49. Gupta, P., Yadav, D. K., Siripurapu, K. B., Palit, G., & Maurya, R., (2007). Constituents of *Ocimum sanctum* with anti-stress activity. *Journal of Natural Products, 70*(9), 1410–1416.

50. Babu, K. N., Hemalatha, R., Satyanarayana, U., Shujauddin, M., Himaja, N., Bhaskarachary, K., & Kumar, B. D., (2018). Phytochemicals, polyphenols, prebiotic effect of *Ocimum sanctum, Zingiber officinale, Piper nigrum* extracts. *Journal of Herbal Medicine, 13*, 42–51.

51. Kusindarta, D. L., Wihadmadyatami, H., Jadi, A. R., Karnati, S., Lochnit, G., Hening, P., Haryanto, A., Auriva, M. B., & Purwaningrum, M., (2018). Ethanolic extract *Ocimum sanctum*. Enhances cognitive ability from young adulthood to middle-aged mediated by increasing choline acetyltransferase activity in rat model. *Research in Veterinary Science, 118*, 431–438.

52. Park, D., Yang, Y. H., Bae, D. K., Lee, S. H., Yang, G., Kyung, J., Kim, D., Choi, E. K., Lee, S. W., Kim, G. H., & Hong, J. T., (2013). Improvement of cognitive function and physical activity of aging mice by human neural stem cells over-expressing choline acetyltransferase. *Neurobiology of Aging, 34*(11), 2639–2646.

53. Shen, X., Xie, Y. Y., Chen, C., & Wang, X. P., (2015). Effects of electroacupuncture on cognitive function in rats with Parkinson's disease. *International Journal of Physiology, Pathophysiology and Pharmacology, 7*(3), 145.

54. Singh, V. K., Das, S., Dwivedy, A. K., Rathore, R., & Dubey, N. K., (2019). Assessment of chemically characterized nanoencapuslated *Ocimum sanctum* essential oil against aflatoxigenic fungi contaminating herbal raw materials and its novel mode of action as methyglyoxal inhibitor. *Postharvest Biology and Technology, 153*, 87–95.

55. Sood, R., & Chopra, D. S., (2018). Optimization of reaction conditions to fabricate *Ocimum sanctum* synthesized silver nanoparticles and its application to nano-gel systems for burn wounds. *Materials Science and Engineering: C, 92*, 575–589.

CHAPTER 7

Anti-Proliferative Potentials of Silver Nanoparticles Synthesized from Natural Biomass

ROSHAN TELRANDHE,[1] SANJAY KUMAR BHARTI,[2] and
DEBARSHI KAR MAHAPATRA[3]

[1]*Department of Pharmaceutics, Smt. Kishoritai Bhoyar College of
Pharmacy, New Kamptee, Nagpur – 441002, Maharashtra, India*

[2]*Institute of Pharmaceutical Sciences, Guru Ghasidas Vishwavidyalaya
(A Central University), Bilaspur – 495006, Chhattisgarh, India*

[3]*Department of Pharmaceutical Chemistry, Dadasaheb Balpande
College of Pharmacy, Nagpur – 440037, Maharashtra, India, E-mail:
mahapatradebarshi@gmail.com*

ABSTRACT

Cancer is a category of diseases within which a cell or a bunch of cells displays uncontrolled growth, invasion, and metastasis. Therapy is the use of anti-tumor medicine to treat cancer by busy bodied the expansion ability of cancer cells (CRC). Metal nanoparticles (NPs) have tremendous applications within the space of chemical change, opto-physical science, diagnostic biological probes, etc. Silver (Ag) could be a metallic element and it is applicable in medicines is priceless for millenniums. The preparation of NPs involves main three strategies: physical, chemical, and biological. As there are varying strategies like sol-gel method, chemical precipitation, reverse particle, hydrothermal methodology, and biological strategies, etc. are employed to synthesize silver nanoparticles (AgNPs). The biologically synthesize AgNPs in nanobiotechnology space have enhanced its importance to make eco-friendly; value effective, stable NPs, and their advantages in medicines, agriculture, and physics. AgNPs are synthesized from completely diverse safe plants and can be applied in pharmaceutical and biological fields. The biological strategies are eco-friendly, value-effective, and do not involve the utilization of harmful chemicals. This book chapter focuses on the synthesis and characterization of AgNPs fabricated from natural plant and organism extracts for the treatment of cancer.

7.1 INTRODUCTION

Cancer is a category of diseases within which a cell or a bunch of cells displays uncontrolled growth, invasion, and metastasis [1]. Unfortunately, at present, there are only cancer therapeutic agents that have limited effect on the host cells; particularly, the bone marrow, solid tissues, reticule epithelium system, and gonads [2]. As a result of the high death rate related to intense cancer therapy and radiation, several cancer patients requested alternative and/or complementary strategies of the treatment. Metal nanoparticles (NPs) have tremendous applications within the space of chemical change, opto-physical science, diagnostic biological probes, etc. These bi-metal NPs principally originate from philanthropist metals like gold, silver, noble metal, and lead. Among these philanthropist metals, silver (Ag) is the metal with wide applications in biological systems, living organisms, and medicine [3]. Many attempted to employ AgNPs as an anti-tumor agent and that they have all turned up positive [4]. It has been stressed over the years that the size reduction of NPs plays a pivotal role in the bioaccessibility which is compatibility for medical specialty applications like cancer [5]. Many plant-derived products are rumored to exhibit potent anti-tumor activity against many eutherian mammals and human neoplastic cell lines.

Nano-product engineering deals with the understanding and regulation matter at a dimension of roughly one to one hundred nanometers. It includes the understanding of the elemental physics, chemistry, biology, and technology of metric linear unit scale objects. It additionally includes areas of computation, sensors, nanostructure materials, biomedical, agricultural, cancer, biotechnology, and miscellaneous areas [6]. Since the late 90s, the AgNPs were synthesized by chemical, physical, and chemical methods [7]. Numerous standard strategies are utilized for the synthesis of NPs. Before adopting these strategies for application in therapeutic purposes, certain components like safety, eco-friendly, and economically viable must be taken into consideration. There are two principle strategies utilized for the NPs synthesis; top down-method and bottom-up method. In the high down method, bulk materials are softened into little particles at the nanoscale with numerous techniques like grinding, edge; which means the NPs units were created by size reduction from a beginning material [8, 9]. In the bottom-up method, the NPs are designed by connexion atoms, molecules, and smaller particles [10] (Figure 7.1).

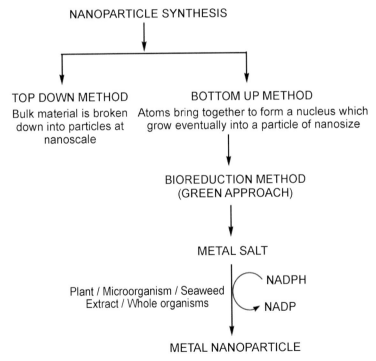

FIGURE 7.1   Top-down and bottom-up methods for AgNP fabrication.

7.2   PREPARATION OF NANOPARTICLE (NP) [11]

The preparation of NPs involves main three strategies: (a) physical, (b) chemical, and (c) biological. Traditionally, NPs were created solely by physical and chemical strategies. Some common strategies involve unit particle sputtering, solvothermal synthesis, reduction, and sol-gel technique (Figure 7.2) [12]. The necessity for the biogenesis of NPs rose because the physical and chemical processes were price-effective. Nature has provided numerous processes for the synthesis of NPs and small length scaled inorganic materials. It has contributed to the event of comparable new materials and for the most part, unknown space of analysis supported the biogenesis of nanomaterials. Hence, for the biogenesis of NPs, plants, and microorganism sources were used. Various strategies return beneath physical and chemical processes like high energy ball edge, soften commixture, physical vapor deposition, optical maser ablation, sputter deposition, mixed route, sol-gel methodology, etc. have been into applications. Among all some strategies that unremarkably employed in the physical and chemical processes, involve:

1. A sol-gel technique that may be a wet chemical technique, wherever two forms of materials or elements, sol, and gel. This methodology is employed for the fabrication of metal oxides from a chemical resolution that acts as a precursor for an integrated network (gel) of distinct particles or polymers. The precursor sol will either be deposited on the substrate to make a film, forged into acceptable instrumentation with the desired form, or into synthesized powders.

2. Solvothermal synthesis may be a versatile vasoconstrictive route within which polar solvents fraught and at temperatures on top of their boiling points. Under solvothermal conditions, the solubility of reactants will increase considerably, enabling the reaction to require place at a lower temperature.

3. The chemical reduction is the reduction of ionic salts in an acceptable medium within the presence of wetting agent exploitation reducing agents. A number of the unremarkably used to reduce the metallic element such as borohydride, hydrate, etc.

4. Optical maser ablation is the process of removing material from a solid surface by irradiating with irradiation. At low optical maser flux, the fabric is heated by absorbing optical maser energy and evaporates or sublimates. At higher flux, the fabric is born-again to plasma. By relying upon the optical properties and also the optical

maser wavelength, the optical maser energy is absorbed and also the quantity of fabric is removed.

5. Gas condensation, wherever totally different metals area unit gaseous in separate crucibles within ultra-high chamber crammed with argon or noble gas at a typical pressure of few 100 Pascals. As a result of lay atomic collisions with gas atoms in the chamber, the gaseous metal atoms lose their mechanical energy and condense within the sort of little crystals that accumulate on cryogen crammed cold finger-like gold NPs units synthesized from gold wires.

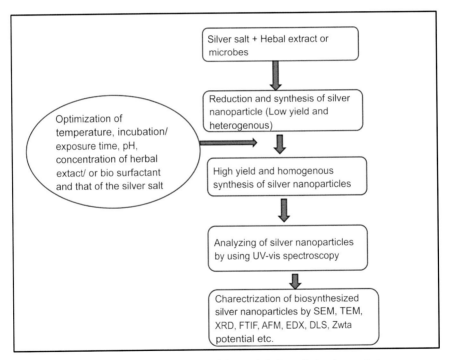

FIGURE 7.2   Synthesis of silver nanoparticles and their optimization techniques.

Characterization of AgNP's physicochemical properties is important for behavior, safety, bio-distribution, and effectiveness of NPs. Therefore, AgNPs characterization is important to gauge the purposeful aspects of synthesized NPs.

7.3 CHARACTERIZATION OF SYNTHESIZED SILVER NANOPARTICLES (AGNPS) IS DONE BY MISTREATMENT NUMEROUS STRATEGIES

7.3.1 *UV-VIS SPECTROSCOPY [13–16]*

UV-VIS analysis is exceptionally valuable and essential for the characterization of NPs. AgNPs have distinctive optical properties that build them firmly get together with specific wavelengths of sunshine (UV/VIS/IR spectrographic analysis). UV-VIS spectrographic analysis is fast, simple, basic, and specific for numerous types of NPs, desires simply a quick amount of your time of estimation. In AgNPs, the conductivity band and valence band lie close to one another within which electrons move brazenly. These free electrons supply ascent to a surface plasmon reverberation (SPR) assimilation band, this can be occurring as a result of the mixture swaying of electrons of AgNPs in reverberation with the sunshine wave. The assimilation of AgNPs depends upon the molecule estimate, nonconductor medium, and artificial surroundings. Observation of this peak assigned to a surface plasmon is incredibly a lot of records for various metal NPs with sizes running from 2 nm to 100 nm.

7.3.2 *X-RAY DIFFRACTION (XRD) [16–18]*

X-ray diffraction (XRD) may be a common analytical technique that has been used for the examination of each atomic and crystal structures, qualitative identification of varied compounds, activity the degree of crystallinity, quantitative resolution of chemical species, particle sizes, isomorphic substitutions, etc. At the purpose once X-ray lightweight reflects on any crystal, it prompts the event of various optical phenomenon styles, and a pattern reflects physicochemical attributes of crystal structures. During a powder sample, diffracted patterns usually originate from the specimen and replicate its structural options. During this means, XRD will examine the essential options of an in-depth sort of materials, as an example, inorganic impetuses, superconductors, biomolecules, glasses, polymers, etc.

7.3.3   FOURIER TRANSFORM INFRARED (FTIR) SPECTROSCOPY [19–21]

FTIR will offer exactitude, duplicability, and what is a more perfect signal-to-noise quantitative relation. By utilizing FTIR spectrographic analysis, it becomes attainable to spot very little absorbance changes that performs distinction spectrographic analysis, wherever one might acknowledge the insufficient assimilation teams of many dynamic deposits from the intensive foundation of the total supermolecule. FTIR spectrographic analysis is commonly accustomed to seeing if biomolecules are related to the merger of NPs that is a lot of articulate in the erudite and fashionable analysis. Besides, FTIR is boot stretched to investigate the nanoscale materials, as an example, the affirmation of helpful atoms covalent united onto silver, carbon nanotubes, graphene, and gold NPs, or co-operation amongst catalyst and substrate amid the chemical procedure. FTIR is an associate degree applicable, important, non-invasive, price-effective, and basic strategy to determine the role of biological molecules within the reduction of silver in its elemental form.

7.3.4   TRANSMISSION MICROSCOPY (TEM) [21]

TEM may be an important, oftentimes used, and demanding system for the characterization of nanomaterials. It is accustomed to getting quantitative measures of the molecule and to boot grain size, size distribution, and morphology. The magnification of TEM is primarily controlled by the quantitative relation of the distance between the target lens and the sample. Therefore, the distance between the objective lens and its image plane is measured.

7.3.5   SCANNING MICROSCOPY (SEM) [21–23]

The field of engineering has given the main thrust within the improvement of various high-determination research procedures with a particular goal that required the nanomaterials by utilizing a light-weight electron emission to probe objects on a fine scale. Amongst completely different microscopy, SEM may be a surface imaging technique, utterly equipped for resolution numerous molecule sizes, size distributions, nanomaterials shapes, and therefore, the surface morphology of the particles at the little scale and nanoscale. Utilizing SEM, the morphology of particles can be acquired and checking the particles physically, or by utilizing specific programming.

7.3.6 ATOMIC FORCE RESEARCH (AFM) MICROSCOPY [22, 23]

AFM is employed to research the aggregation and dispersion of nano-materials, additionally to their size, shape, sorption, and structure. Three completely different scanning modes are obtainable, together with contact mode, non-contact mode, and intermittent sample contact mode. It also can be used to characterize the nanomaterial's interaction with supported lipid bilayer in real-time that is not realizable with current microscopy techniques. AFM does not need oxide-free, electrically semi-conductive surfaces for the measure, does not considerable harm too many sorts of native surfaces, and it will answer to the sub-nanometer scales in binary compound fluids.

7.4 ANTI-CANCER ACTIVITY OF SILVER NANOPARTICLES (AGNPS) [24, 25]

Cancer is one in all the deadly diseases that have infected developed as well as developing countries. Investigated the molecular mechanism of AgNPs (Figure 7.3) and located that programmed death was concen-tration-dependent; there is a demand to develop them to scale up the general effects. Further, the synergistic result of cell death mistreatment U-phosphoribosyl transferase expressing cells and non-U-phosphoribosyl transferase expressing cells within the presence of anti-metabolite enhanced the hope of its use as nanomedicine. In these conditions, it was determined that AgNPs not solely induce cell death, however, conjointly sensitize cancer cells (CRC) and conjointly rumored that silver embedded magnetic NPs showed significant activity against breast-CRC and floating malignant neoplastic disease cells. The plant extract mediated synthesized AgNPs demonstrated cytotoxic result against the human respiratory organ CRC (A549) that indicated AgNPs might target cell-specific toxicity.

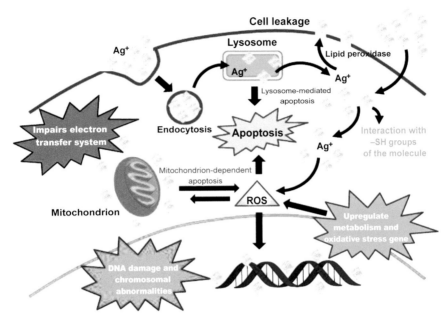

FIGURE 7.3   Anti-cancer mechanism of action(s) of AgNPs.

7.5   NANOPARTICLES (NPS) FOR CANCER THERAPY [26–31]

Cancer is one in all the foremost common issues and high health issues during this world. It has been determined that a disease focus can develop some varieties of cancers in their entire period. Based on the origin, there are types of cancer that exist like the thyroid, prostate, bladder cancer, urinary organ cancer, pancreatic, carcinoma, melanoma, a malignant neoplastic disease with every kind, carcinoma, colon-rectal combined cancer, etc. In cancer, the cells divide and grow uncontrollably, forming malignant tumors and invasive near elements of the body. To date, a whole cure for this current illness is yet to be discovered. For the last three decades, researchers, and clinicians have developed varied strategies to either fore-stall or inhibit the expansion of cancer. The most common styles of cancer treatment are surgery, chemotherapy, actinotherapy, and photodynamic medical care. Surgery is the most prevalent method to treat and diagnose localized cancer. In the case of the tumors, significantly once cancer has not metastasized to body fluid nodes or a different part of the body, actinotherapy is employed. The approach involves high energy radiation to kill CRC by damaging their deoxyribonucleic acid. Actinotherapy is provided along with surgery or together with different treatment regimens (chemotherapy). Though actinotherapy is taken into account as a widely accepted model of cancer treatment, however, it might super-sensitized each cancer and traditional death.

Therapy is the use of anti-tumor medicine to treat cancer by busy bodied the expansion ability of CRC. Completely different chemotherapeutic medicine will be utilized in a prescribed dose in numerous sorts of cancer that targets explicit proliferation pathways. Immunotherapy is a treatment that uses the host own system to assist fight cancer. Photodynamic medical care could be a treatment that uses special medicine, referred to as photo-sensitizing agents, together with light-weight to kill CRC. Supported the chemical structure and mechanism by that they act, the therapy medicine will be divided into four groups:

1. Alkalyting agents that harm deoxyribonucleic acid to stop the expansion of CRC;
2. Opposed metabolites that interfere with the replication of deoxy-ribonucleic acid or transcription of the polymer by work the traditional building block of polymer and deoxyribonucleic acid and henceforth will cause regular cell cycle arrest;
3. Anti-tumor antibiotics that embrace anthracyclines, actinomycin-D; and
4. Mitotic inhibitors are principally plant-originated compound derived from a natural product (NP) that stop the cellular division and henceforward inhibits the expansion of the cell cycle.

As therapy might reach all the parts together with CRC, there is also the risk prevalence of some aspects throughout the treatment. To avoid this, recently, scientists have discovered NPs may be used as a targeted drug delivery system for cancer medical care wherever solely the CRC can destroy while not poignant healthy traditional cells. Due to their large applications within the medical specialty space, these are getting used as carriers for hydrophobic medicine, diagnostic, and therapeutic func-tions. Additionally, to the present, they will even be scaled back due to the toxicity of a therapeutic drug. As NPs are utilized in varied functions, however, their use in anti-neoplastic delivery, it is an additional result of the medicine that is simply soluble and can penetrate deep in organs and tissues. Also, these are utilized in fluorescent biological labels (diagnostic purpose), drug, and sequence delivery, detection of pathogens, detection of proteins, inquiring of deoxyribonucleic acid structure, tissue engineering, growth damaging via heating (hyperthermia), separation, and purification of biological molecules and cells, magnetic resonance imaging distinction sweetening, phagokinetic studies, etc.

Silver (Ag) could be a metallic element and it is applicable in medicines is priceless for millenniums. As there are varying strategies like sol-gel method, chemical precipitation, reverse particle, hydrothermal methodology, and biological strategies, etc. are employed to synthesize AgNPs. The biological strategies are eco-friendly, value-effective, and do not involve the utilization of harmful chemicals. AgNPs are synthesized from completely diverse safe plants and can be applied in pharmaceutical and biological fields.

7.6 PLANT AND ORGANISM MEDIATED FABRICATED SILVER NANOPARTICLES (AGNPS) AS ANTI-CANCER THERAPEUTICS [29–34]

The rapid synthesis methodology of AgNPs from the fruits, leaves, seeds, and root extract of genus *Citrullus colosynthis* was evaluated against four different human neoplastic cell lines; MCF-7 (breast carcinomas), HepG2 (hepatocellular carcinomas), Caco-2, and HCT-116 (colon carcinomas). The characterization performed through UV-Vis spectroscopic analysis and TEM analysis showed that the produced NPs of an irregular shape with a median size of 562.4 nm. However, the common mean size of AgNPs of totally different components of *C. colocynthis* such as fruits, leaves, seeds, and roots was 267 nm, 16.578 nm, 13.376 nm, and 7.398 nm, respectively. The cell viability assay of the AgNPs on human CRC showed that HepG2 cell line and HCT-116 cell line were the foremost sensitive cell line with IC_{50} values of 21.2 µg/ml, 22.4 µg/ml, respectively, whereas the caco-2 cell line was the foremost resistant cell line towards the cytotoxic activity.

AgNPs fabricated with oregano leaf extracts were tested against carcinoma cell A549, where IC_{50} was achieved at a 100 µg/ml. UV-Vis spectroscopic analysis presented AgNPs morphology in the range of 63–85 nm.

Similarly, screening of AgNPs of size 5–30 nm (characterized by UV-Vis spectral analysis and scanning electron microscope), produced from *Cissus quadrangularis* extracts against two human carcinoma cell lines; MCF-7 and HepG2 presented IC_{50} of 20 µg/ml and 64 µg/ml, respectively.

The liquid extract of marine micro-algae *Ulva asterid* dicot genus was applied to synthesize AgNPs and screened concurrently against human neoplastic cell lines Hep-2, MCF-7, and HT-29 because of its earlier reported antibacterial drug, antiviral, and metastatic tumor activity. Analysis through UV-Vis spectroscopic, Fourier transform infrared (FTIR), and XRD revealed size of 20–56 nm. The anti-proliferative study highlighted remarkable activity with IC_{50} values of 12.5 µg/ml, 37 µg/ml, and 49 µg/ml, respectively.

The event of safe and eco-friendly method for the synthesis of silver nanoparticles (AgNPs) from *Dilleniid dicot* genus oleracea was performed to visualize the metastatic tumor result against MCF-7. The total cauliflower florets fabricated AgNPs of size 48 nm, as characterized by UV-Vis spectroscopic analysis, FTIR, XRD, and SEM. These AgNPs were found to inhibit the proliferation of MCF-7 cell with IC50 values of 190.501 µg/ml.

Liquid extracts of algae *Gelidiella* sp. were applied to synthesize AgNPs of size 40–50 nm as characterized by sophisticated analytical instruments. The cytotoxic activity against human neoplastic cell line HepG2 showed IC_{50} of 31.5 µg/ml.

From Table 7.1, it may be ascertained that SNPs from liquid extracts of fruits and roots of *C. colosynthis* and algae *Ulva asterid* dicot genus extracts against HepG2 square measures the foremost effective in terms of metastatic tumor activity in breast and cancer of the liver cells. But the liquid extracts from oregano and *Cissus quadrangularis* plant shows non-significant result against neoplastic cell lines.

TABLE 7.1   List of Plants Mediated Synthesized AgNPs Having Reported Anti-Cancer Activity

Plants	Source for AgNPs Synthesis	Human Cancer Cell Line(s)	Size of AgNPs (nm)	IC_{50} (µg/ml)
Brassica oleracea [33]	Cauliflower florets	MCF-7	48	190.5
Seaweed Gelidiella sp [34]	Whole seaweed	HepG2	31.25	40.5

TABLE 7.1 *(Continued)*

Plants	Source for AgNPs Synthesis	Human Cancer Cell Line(s)	Size of AgNPs (nm)	IC$_{50}$ (µg/ml)
Piper nigrum [35]	Fruits	MCF-7 and Hep-2	20	MCF-7 = 52.7 and HepG2 = 43.3
Citrullus colosynthis [36]	Roots, Fruits, and Leaves	HCT-116, MCF-7, and HepG2	7.398, 19.267, 13.376	HCT116 = 21.2; MCF-7= 22.4; and HepG2 = 22.4
Origanum vulgare [37]	Leaves	A549	63–85	100.2
Cissus quadrangularis [38]	Stem	Hep-2	20–56	64.9
Seaweed Ulva Lactuca [39]	Whole microalgae	Hep-2, MCF-7, and HT-29	5–30	HepG2 = 12.5; MCF-7 = 37.8; and HT-29 = 49.6

7.7 CONCLUSION

The biologically synthesize AgNPs in nanobiotechnology space have enhanced its importance to make eco-friendly; value effective, stable NPs, and their advantages in medicines, agriculture, and physics. From selection analysis on applied natural sciences for AgNPs, it is found that the components are safer and therapeutically active. These pharmaco-therapeutically privileged AgNPs are recently gaining a lot of attention due to their potential to treat the majority of diseases. However, many issues associated with AgNPs like poor solubility, poor bioavailability, low oral absorption, instability, and unpredictable toxicity of limited their use. Hence, different NPs together with chemical compounds, liposomes, proliposomes, solid-lipid NPs, and microemulsions utilization have shown the potential to deliver medicines with higher cancer medical aid.

KEYWORDS

- **anti-cancer**
- **atomic force research**
- **Fourier transform infrared**
- **green synthesis**
- **nanoparticle**
- **X-ray diffraction**

REFERENCES

1. Kundu, S. S., Gupta, M., Mazumder, U. K., Haldar, P. K., Saha, P., & Bala, A., (2011). Antitumor activity of citrus maxima (Burm.) Merr. leaves in EHRLICH'S ascites carcinoma cell-treated mice. *ISRN Pharmacology, 1*, 1–4.
2. Nitha, B., Meera, C. R., & Janardhanan, K. K., (2005). Antitumor activity of ethanolic extract of *Lentinus dicholamellatus. Amala. Res. Bull., 25*, 165–168.
3. Parashar, V., Parashar, R., Sharma, B., & Pandey, A. C., (2009). Parthenium leaf extract mediated synthesis of silver nano particles: A novel approach towards weed utilization. *Digest Journal of Nanomaterials and Biostructures, 4,* 45–50.
4. Vaidyanathan, R., Kalishwaralal, K., Gopalram, S., & Gurunathan, S., (2009). Nano silver the burgeoning therapeutic molecule and its green synthesis. *Biotechnology Adv., 27*(6), 924–937.
5. Kim, J. S., Kuk, E., Nam, K., Kim, J. H., Park, S. J., Leo, H. J., et al., (2007). Antimicrobial effect of silver nanoparticles. *Nanomedicine, 3*, 90–101.
6. Albert, M. A., Evans, C. W., & Ratson, C. L., (2006). Green chemistry and the health implications of nanoparticles. *Green Chem., 8,* 417–432.
7. Telrandhe, R., Mahapatra, D. K., & Kamble, M. A., (2017). *Bombax ceiba* thorn extract mediated synthesis of silver nanoparticles: Evaluation of anti-*Staphylococcus aureus* activity. *Int. J. Pharm. Drug Analysis, 5*(9), 376–379.
8. Meyers, M. A., Mishra, A., & Benson, D. J., (2006). Mechanical properties of nanocrystalline materials. *Prog. Mater. Sci., 51*, 427–556.
9. Telrandhe R., (2016). Nanotechnology for cancer therapy: Recent developments. *Eur. J. Pharm. Med. Res., 3*(11), 284–294.
10. Hutchison, J. E., (2008). Greener nano science: A proactive approach to advancing applications and reducing implications of nanotechnology. *ACS Nano, 2*, 395–402.
11. Mohanpuria, P., Rana, N. K., & Yadav, S. K., (2008). Biosynthesis of nanoparticles: Technological concepts and future applications. *J. Nanopart. Res., 10*, 507–517.
12. Siddiq, K. S., Husen, A., & Rao, R., A. K., (2018). A review on biosynthesis of silver nanoparticles and their biocidal properties. *J. of Nanobiotechnology, 16*(14), 1–28.

13. Sondi, I., & Salopek-Sondi, B., (2004). Silver nano particles as antimicrobial agent: A case study on *E. coli* as a model for gram-negative bacteria. *J. Colloid Interface Sci., 275*, 177–182.

14. Zhang, X. F., Liu, Z. G., Shen, W., et al., (2016). Silver nanoparticles: Synthesis, characterization, properties, applications, and therapeutic approaches. In: Yan, B., (ed.), *Int. J. of Mol. Sci., 17*(9), 1534.

15. Stephan, T. S., Scott, E. M., Anil, K. P., et al., (2006). Preclinical characterization of engineered nano particles intended for cancer therapeutics. In: Amiji M. M., (ed.), *Nanotechnology for Cancer Therapy* (pp. 105–137). CRC Press; Boca Raton, FL, USA.

16. Sadeghi, B., & Gholamhoseinpoor, F., (2015). A study on the stability and green synthesis of silver nanoparticles using *Ziziphora tenuior* (Zt) extract at room temperature. *Spectrochim Acta Part A: Mol Biomol Spectrosc., 134*, 310–315.

17. Robin, T. M., (2009). Introduction to powder diffraction and its application to nanoscale and heterogeneous materials. *Nanotechnol. Undergrad. Educ., 1010*, 75–86.

18. Santhoshkumar, T., Rahuman, A. A., Rajakumar, G., et al., (2011). Synthesis of silver nanoparticles using *Nelumbo nucifera* leaf extract and its larvicidal activity against malaria and filariasis vectors. *Parasitol Res., 108*, 693–702.

19. Kim, S., & Barry, B. A., (2001). Reaction-induced FT-IR spectroscopic studies of biological energy conversion in oxygenic photosynthesis and transport. *J. Phys. Chem. B., 105*, 4072–4083.

20. Gurunathan, S., Han, J. W., Kwon, D. N., & Kim, J. H., (2014). Enhanced antibacterial and anti-biofilm activities of silver nanoparticles against gram-negative and gram-positive bacteria. *Nanoscale Res. Lett., 9*(1), 373.

21. Lin, P. C., Lin, S., Wang, P. C., et al., (2014). Techniques for physicochemical characterization of nanomaterials. *Biotechnol Adv., 32*(4), 711–726.

22. Xiang, D. X., Chen, Q., Pang, L., et al., (2011). Inhibitory effects of silver nano particles on H1N1 influenza A virus *in vitro. J. Virol. Methods., 178*, 137–142.

23. Yao, H., & Kimura, K., (2007). Field emission scanning electron microscopy for structural characterization of 3D gold nanoparticle superlattices. In: Méndez-Vilas, A., & Díaz, J., (eds.), *Modern Research and Educational Topics in Microscopy* (pp. 568–575). Formatex Research Center; Badajoz, Spain.

24. Gopinath, P., Gogoi, S. K., Chattopadhyay, A., et al., (2008). Implications of silver nano particle-induced cell apoptosis for in vitro gene therapy. *Nanotechnology, 19*(7), 075104.

25. Gurunathan, S., Jeong, J. K., Han, J. W., et al., (2015). Multidimensional effects of biologically synthesized silver nanoparticles in *Helicobacter pylori, Helicobacter felis,* and human lung (L132) and lung carcinoma A549 cells. *Nanoscale Res. Lett., 10*, 10–17.

26. Li, S., Shen, Y., Xie, A., Yu, X., Qiu, L., Zhang, L., & Zhang, Q., (2007). Green synthesis of silver nanoparticles using *Capsicum annuum* L. extract. *Green Chem., 9*, 852–858.

27. Shankar, S. S., Rai, A., Ahmed, A., & Sastry, M., (2004). Rapid synthesis of Au, Ag and bimetallic Au core-Ag shell nanoparticles using NEEM (*Azadirachta indica*) leaf broth. *J Colloid Interface Sci., 275*, 496–502.

28. Murthy, Y., Kondala, R. T., & Singh, R., (2010). Synthesis and characterization of nano silver ferrite composite. *Journal of Magnetism and Magnetic Materials, 322,* 2071–2074.

29. Panacek, A., Kvitek, L., Prucek, R., Kolar, M., Vecerova, R., Pizurova, N., Sharma, V. K., Nevecna, T. J., & Zboril, R., (2006). *The Journal of Physical Chemistry B., 110,* 16248–16253.

30. Sharma, V. K., Yngard, R. A., & Lin, Y., (2009). *Advances in Colloid and Interface Science, 145,* 83–96.

31. Gardea-Torresdey, J. L., Gomez, E., Peralta-Videa, J. R., Parsons, J. G., Troiani, H., & Jose-Yacaman, M., (2003). Alfalfa sprouts: A natural source for the synthesis of silver nanoparticles. *Langmuir, 19,* 1357–1361.

32. Kanchana, A., & Balakrishna, M., (2011). Anti-cancer effect of saponins isolated from *Solanum trilobatum* leaf extract and induction of apoptosis in human larynx cancer cell lines. *Int. J. of Pharm. and Pharmaceu. Sci., 3,* 356–364.

33. Ranjitham, M. A., Suja, R., Caroling, G., & Tiwari, S., (2013). *In vitro* evaluation of anti oxidant, antimicrobial, anticancer activities, and characterization of *Brassica oleracea. Var bortrytis* L. synthesized silver nanoparticles. *Int. J. of Pharm. and Pharmaceutical Sci., 5*(4), 239–251.

34. Devi, J. S., Bhimba, B. V., & Ratnam, K., (2012). *In vitro* anti cancer activity of silver nanoparticles synthesized using the extract of *Gelidiella* sp. *Int. J. of Pharm. and Pharmaceutical Sci., 4*(4), 710–715.

35. Krishnan, V., Bupesh, G., Manikandan, E., Thanigai, A. K., Magesh, S., Kalyanaraman, R., & Maaza, (2016). Green synthesis of silver nanoparticles using *Piper nigrum* concoction and its anticancer activity against MCF-7 and Hep-2 cell lines. *J. of Antimicrobial Agents, 2*(3), 1–5.

36. Shawkey, A. M., Rabeh, M. A., Abdulall, A. K., & Abdellatif, A. O., (2013). Green nanotechnology: Anticancer activity of silver nano particles using *Citrullus colocynthis* aqueous extracts. *Adv. in Life Sci. and Tech., 13,* 60–70.

37. Renu, S., Arunachalam, K., Annamalai, P., Selvaraju, K., Kanchi, S. S., & Vilwanathan, R., (2013). *Origanum vulgare* mediated biosynthesis of silver nano particles for its antibacterial and anticancer activity. *Colloids and Surfaces B: Biointerfaces, 108,* 80–84.

38. Renugadevik, K., Inbakandan, D., Bavanilatha, M., & Poornima, V., (2010). Cissus quadrangular is assisted biosynthesis of silver nanoparticles with antimicrobial and anticancer potentials. *Int. J. Pharm. Bio. Sci. x, 3*(3), 437–445.

39. Devi, J. S., & Bhima, B. V., (2012). *Anticancer Activity of Silver Nanoparticles Synthesized by the Seaweed Ulva Lactuca In Vitro, 1,* 242. doi: 10.4172/scientificreports.242.

CHAPTER 8

Nanoparticles as a Promising Technology in Microbial Pharmaceutics

EKNATH D. AHIRE[1], SWATI G. TALELE[2], and HEENA S. SHAH[1]

[1]Department of Pharmaceutics, SSS's, Divine College of Pharmacy, Nampur Road, Satana, Maharashtra, India

[2]Department of Pharmaceutics, Sandip Institute of Pharmaceutical Sciences, Mahiravani, Nashik, Maharashtra, India,
E-mail: swatitalele77@gmail.com

*Corresponding author. E-mail:

ABSTRACT

Antimicrobial resistance is a budding problem that has obstructed the world of nearby starting of the end for the old generation antimicrobials. Nanoparticles give the positive response to anti-microbial resistance, which could trigger the innovation and makes new generation of anti-microbials for the treatment in future using nanotechnology based drug delivery. Nanotechnology is progressively exhausting technology both in nano-medicine and nano-material for the diagnosing and treatment of innumerable diseases and disorders. Nanoparticles presenting worthy potential in treatment of bacterial infection, where developed several drug deliveries like microbial triggered drug delivery, micro-encapsulation, pH sensitive, surface charge changing nanoparticles for treating various bacterial infection. It will be possible by developing novel formulation into nanotechnologies can increase therapeutically longest duration to act against microbial infection. However, many unique physicochemical characteristics of nanoparticles can offer new antibacterial modes of action which can be explored. Antimicrobial nanoparticles and nanonized drug delivery transporters have arisen as effective agents against various microbial infections. Nanoparticles having distinctive properties other than their small and tailored particle size like improved reactivity, functionalizable structure and appreciable surface area. In contrast to the conventional antimicrobial agents, nanoparticles help in reducing the toxicity, controlling resistance and reducing the cost. Additionally, nanoparticles are also improving the therapeutic and pharmacokinetic properties of the antimicrobial agents. Nanotechnology also helps in development of cost effective, accurate and fast diagnosis, detection and management by treating microbial agents. They were working as nano-anti microbial agents and their extensive ability of treating diseases and disorders. Besides, nanoparticles use multiple biological pathways to apply their antimicrobial mechanisms like destruction of cell wall, DNA inhibition, and enzyme synthesis. Furthermore, the preparation of these nanoparticles is more cost effective than antibiotics synthesis and also more stable for long term stability and they are also withstand to grating condition. In present chapter, we are focusing upon the nanoparticles based drug delivery system as promising technology in microbial pharmaceutics and extensive applications thereof.

8.1 INTRODUCTION

Nanotechnology based drug delivery system is appropriate drug transporter or vehicle for overwhelming pharmacokinetic restrictions related with conventional drug delivery. Nanoparticles, like silica nanoparticles [1], carbon nanotubes [2], nano complexes [2], liposomes etc. have established benefits of therapeutic action as antimicrobial drug delivery. Nanoparticles based drug delivery is exploratory technology in several diverse diseases used as antimicrobial drug therapy [3]. Antimicrobial nanoparticles generally made up of metals and many biological derived materials [1]. So far, recent several reports have established progresses in discerning targeting of bacterial membrane for lysis, enhanced drug delivery, improved drug function also the ability for targeted accumulation at site of infection because of enhanced vascular permeability [4]. The precise mechanism of nanoparticle action beside different bacteria are not known completely. Nanoparticles are capable to attach with the microbial cell membrane due to electrostatic charge and disturb the integrity of microbial membrane. Nanoparticle toxicity is usually generated by the stimulation of oxidative stress due to free radicals development [5]. Recently, researchers published lots of data related to the uses of nanoparticles in anti-microbial targeting, anti-microbial resistance, different surface charge-substituting polymeric nanoparticles for microbial cell wall-targeted delivery [6], silver nanoparticles as promising approach against multidrug resistant, biosynthesis of nanoparticles by microorganisms and many more [7, 8]. The mechanism of nanoparticles toxicity is depends on the surface manipulations, composition of nanoparticle, basic properties of nanoparticle and the microbial species. Even though there are good relationship in some features of antimicrobial action of nanoparticle. [9]. Biological barriers generates huge difficulty in drug delivery system. Difficulties in nonspecific distribution and insufficient accumulation of drugs, still rests as big tasks to the formulation development scientists. Nanoparticulate based drug delivery provides platform have appeared as appropriate carrier for reducing pharmacokinetic limitations related to conventional drug delivery system. Nanoparticles based delivery gives the best effect to cross the microbial barrier and reached the therapeutic agents at the site of diseased [3]. Exclusively, numerous classes of antimicrobial nanoparticles and different nano-sized carriers for antimicrobial delivery have established their efficiency meant for the treatment of microbial infections as well as MDR [10]. Since, nanoparticles having great surface area to volume ratio and its exclusive physicochemical properties, nano-material is favourable antimicrobial agents of the new era. Nanoparticles are themselves acts as effective antimicrobial agents for different of therapeutic applications, like, nanoparticles based dressings [11], nonogels, nanolotions and nanoparticles coated medical devices [12–14]. Different nanoparticles have been explored as effective antimicrobial delivery carrier and which is also protects the drugs from the resistance mechanism in a targeted microorganism. A nanoparticle provides several potentially synergistic and independent methods on the same way, to increase antimicrobial activity and overwhelms the MDR. At this stage, there very less reported literature regarding toxicity and clinical applications of nanoparticles as antimicrobials themselves and as carrier of antimicrobial agents [15]. Though, the most significant scientific progress on development of nano-sized vesicles with distinctive biochemical and physical properties for drug delivery applications have taken place within last few decades. The common examples are liposomes, dendrimers, polymeric nanoparticles [16], lipid based nanoparticles, virus based nanoparticles, and lipid based nanoparticles and many more as shown in Figure 8.1. Nanoparticles have the capability to increase therapeutic index of presently available drugs by reducing drug toxicity, increasing efficacy of drugs, and reaching the steady state therapeutics over a extended period of time. Nanoparticles also enhanced the drug stability as well as solubility, for the development of potentially active new chemical moieties. The flexibility in surface chemistry of nanoparticles also permits for probable conjugation of targeting ligands for targeted drug delivery. Nevertheless, many challenges rests for nano medicine based drug delivery to counteract against MDR. Drug loaded nanoparticles presently have the opportunity to distribute within the normal tissue [17–19].

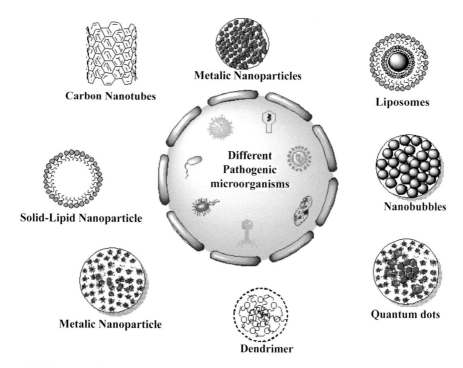

FIGURE 8.1   Various nanoparticulate drug delivery systems to treat infections or drug resistance caused by different pathogenic microorganisms.

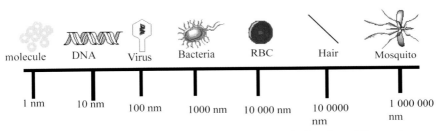

| molecule | DNA | Virus | Bacteria | RBC | Hair | Mosquito |

| 1 nm | 10 nm | 100 nm | 1000 nm | 10 000 nm | 10 0000 nm | 1 000 000 nm |

FIGURE 8.2 Nanoscale for the understanding of nanoparticulate drug delivery size.

8.2 HISTORY

Overlast few decades' biopharmaceutical field facing the difficulties of antimicrobial drug resistance similarly known as multi drug resistant (MDR) [20]. Microbial resistance developed to lots of approved active therapeutics are increasingly common and progressively spread worldwide. Opportunistic microorganisms or Pathogenic microbe's considering, this as a golden phase to live and grow in comfortable environment without any uncomfortness. MDR of microbes direct impact on the expressively increase in cost of long period treatment, mortality and morbidity of the patients [8]. Progressively, large number of antibiotics has been prescribed gratuitously and due to unconsciously use of this antibiotics, MDR effects being increased day by day. Infectious diseases is the global problem causes several deaths per years, only due to MDR effect of antimicrobials [21]. Amalgamations of various factors related with biological surroundings along with instability and poor solubility of numerous therapeutically active molecules causes less poor bioavailability and it is requirement of to develop new drug delivery, which can minimize these problems. The structure of cancerous tumors shows lots of challenges for targeting them. Though, it furthermore gives exclusive prospect to focus on the nanotechnology based drug delivery [2]. The common use of antibiotics in humans establishes a significant cause of antibiotic resistance. In large amount of antimicrobials has been used in plants as growth promoters or as pesticide, which also contributes in some amount for antimicrobial resistance [22]. We are at the phase, where we are not capable to surely treat an increasing amount of microbial infection. It is necessary to take strong action against this problem or to find out special innovative and unique drug delivery systems to overcome particular problem [23]. To counteract against this problem we have some options, which have to choose them, as follows: (a) developing the education program for the medical and paramedical practitioner on use of antibiotics; (b) reduce use of antibiotic agents as possible as; and (c) develop such novel drug delivery to overcome MDR problem. First two approaches are depends on public and government interest, but another, third one approach is consider as promising technology, which can be counter the MDR effect [22].

In the present antimicrobial resistance era, we should accept and to face this worldwide challenge to carry on enjoy the outputs of antimicrobial agents and enhanced quality of life. While, the regulatory problem of emerging novel antimicrobial agent are very slowly accepted to the streamline and its very long process. The major scientific problem to categorising and developing new antimicrobial agent remains difficult. From last two decades of research work shows that antimicrobial agents that are very effective, safer dose and broad spectrum antimicrobials are difficult to search [24]. Practically, all microbial infectious agents such as parasites, fungi, bacteria and viruses have active level of resistance and mortality rate, consequently they are known as super bugs. Decreasing efficiency of antibiotics while treating general infections has accelerated within recent few years. We are reached at the stage of untreatable bacterial strains, and we are now at the dawn of post-antibiotic time [25]. In developed and developing countries like India constant high amount of use of antibiotics in community and various hospitals [26]. Resistance increases as a significance of mutation in microorganisms and choice pressure from antibiotic applications that gives the reasonable advantages meant for mutated strains [27]. Redeveloped naturally, it may be cause of poor sanitary condition, incorrect use of antimicrobial agents, underprivileged infection inhibition and unsafe food handling. Bearing in mind the importance of MDR effect, there is need to be understand the problems related with MDR and essential to recognize its importance to prevent microbial resistance by developing novel drug delivery [2, 20]. The nanoscale for the understanding of nanoparticulate drug delivery as shown in Figure 8.2.

8.3   CLASSIFICATION OF NANOPARTICULATE SYSTEMS USED IN MICROBIAL PHARMACEUTICS

NPs are classified in different classes, on the basis of their nature, properties, and production and drug delivery methods as explained in Table 8.1.

TABLE 8.1 Different Types Nanoparticulate Systems

Sl. No	Nanoparticulate System	Particle Size	Applications	Characteristic of Nanoparticulate System	Advantages Over Conventional Delivery	References
1.	Polymeric nanoparticles	10–1000 nm	Good drug carrier for sustained and controlled drug delivery	Biocompatible as well as biodegradable polymer provides total protection	Less toxic Better stability Target specificity	[28, 30]
2.	Metallic nanoparticles	Bellow 100 nm	Gene and drug targeted delivery, DNA fingerprinting, DNA analysis, gene therapy, MRI diagnosis, Guided drug delivery	Silver, gold, alloy, Magnetic are very small in size and provides high surface area for action, self-acts as a microbial toxicant, stable	Surface functionalization possible easily	[31]
3.	Polymeric micelles	10–100 nm	Diagnosis, targeted drug delivery	High drug entrapment, Biostable, copolymerization	Good carrier	[32]
4.	Liposomes	50–100 nm	Provides active and passive Gene, protein, and peptide delivery	Biocompatible, good entrapment efficiency, phospholipid vesicle	Solubility enhancement	[33]
5.	Carbon nanotubes	20–1000 nm in length and 0.5–3 nm in dia.	Cell specificity, Solubility enhancement, gene silencing vaccine, gene, and peptide targeted drug delivery	Good strength and remarkable electrical properties like insulating, conductivity, and semiconductivity	Improved physicochemical properties	[34, 35]
6.	Nanocrystals Quantum dots	2–10 nm	Imaging, Cancer therapy, Immunoassay, In liver diseases diagnosis, as a breast cancer marker	Broad UV light excitation and high photostability	Improved optical properties	[36]

TABLE 8.1 *(Continued)*

Sl. No	Nanoparticulate System	Particle Size	Applications	Characteristic of Nanoparticulate System	Advantages Over Conventional Delivery	References
7.	Dendrimer	Bellow 10 nm	Anti-retroviral therapy, Transfection, a contrast agent for imaging, rapid clearance, Controlled drug delivery, targeted drug delivery for bioactives	Highly branched in structure, containing three parts core, branch, and surface all parts are used for drug loading, long circulatory in shape,	Ultra-lightweight, high surface area, high thermal conductivity	[37]
8.	Paramagnetic nanoparticles	Bellow 100 nm	Cancer therapy, Protein identification, imaging, Targeted action	Conjugation with antibodies, Detect breast cancer cells with MRI,	Targeting	[32]
9.	Nanosomes	Bellow 1000 nm	Brain targeting, tumor-targeting ROS production	Diagnosis, targeting, and therapy	Safer than chemotherapy (without side effect)	[38]
10.	Nanopores	20 nm	DNA sequencing, pharmacogenomics in drug development, diabetes treatment	Consist of wafers with highly dense pores	Doesn't allow entry of oxygen and glucose inside.	[28, 38]
11.	Microbivores	2000 nm	Clear the blood circulation	Greater efficiency in phagocytosis	Blood purification	[39]

TABLE 8.1 *(Continued)*

Sl. No	Nanoparticulate System	Particle Size	Applications	Characteristic of Nanoparticulate System	Advantages Over Conventional Delivery	References
12.	Nanosuspension and nanocrystals	50–1000 nm	Safe and effective passage, favorable drug delivery system	Improved bioavailability	Solubility enhancement of poorly soluble drugs	[28]
13.	Solid-lipid nanoparticles	Bellow 1000 nm	Diverse drug delivery route, non-viral transfection	Controlled drug delivery	Good carrier for drug	[28]
14.	Nanoshells	1–20 nm	Cancer therapy, diagnosis, hydrogel mediated delivery, in diabetes	Core of silica and metallic outer layer	Targeted delivery	[40, 41]
15.	Nanobubbles	Bellow 1000 nm	Delivery of drug, gene therapy, thrombolysis	Stable at room temp	Targeting to the tumor tissue	[42]

NPs are commonly classified in three major classes: (1) one-dimension NPs consist of thin films in size 1–100 nm in ranges. (2) two-dimension NPs, it contains nanotubes and related nanoparticulate systems; and (3) three dimension NPs contains quantum dots, dendrimers, and fullerenes [28, 29].

8.4 NANOPARTICLES (NPS) ADVANTAGES OVER CONVENTIONAL DRUG DELIVERY SYSTEMS

NPs have been reported for their possible use in targeted delivery to enhance the bioavailability, reducing drug toxicity, decreasing side effects to the organs in the body. Existing, nanoparticulate based delivery provides cost-effective as well as practicable in hydrophilic and hydrophobic nature through different routes of administration like oral, intraocular, intranasal, and parental. Many advantages of NPs based drug delivery system over conventional drug delivery as mentioned bellow [28]:

- Nanoparticle has the capability to incorporate, conjugate, or encapsulate various therapeutic agents in relation to target to definite cells or tissues and also provides the controllable drug release in the human body [43].
- NPs have to ability to deliver the drug to minute parts in the body [44].
- Newly designed nanoparticulate based drug delivery provides good control over the physical properties regarding control drug delivery of various drugs using different biomaterials and polymers [45].
- Nano based drug delivery offers to deliver various biologicals, even in a complex part of the body such as blood-brain barrier [46].
- NPs give competent drug delivery to enhance the aqueous solubility of poorly water-soluble drug candidates, which enhances bioavailability-controlled release of drugs and specific targeted delivery [47].
- Nanoparticulate drug delivery system overcomes the MDR effect caused by the different physiological barriers.

- Highly quick identification and quantification of tumor cells by means of NPs based drug delivery [48].
- Drug particle surface characterization and particle size can be easily manipulated to obtain active and passive drug targeting during parenteral administration.
- Targeting ligands on the surface of NPs for site-specific drug targeting can be obtained by the nanoparticulate drug delivery.
- Nanoparticulate drug delivery provides more specific drug delivery, which reduces the drug toxicity by distribution within the unwanted sites.
- Manipulating surface-modified drug delivery avoids immune response during targeting of the biologicals such as protein, peptides, and nucleic acids, etc.
- Controlled and sustained drug delivery can be achieved easily by using nano-based drug delivery. Due to controlled particle size, drug targeting is possible to the specific site of the body.
- NPs are also employed in the different diagnostic tests for the diagnosis of the various diseases and disorders in the body such as cancer biomarkers for rapid identification of cancer cells (CRC) [49, 50].
- NPs fruitfully exploited to develop the effectiveness of fluorescent markers for medical imaging purposes. A fluorescent NP is significantly avoiding the conventional problems and provides the major advantages towards clinical applications.
- Nanotechnology furthermore has several inferences for *in-vivo* diagnostic devices like swallowable imaging pill and in endoscopic instruments [51].
- Nanoparticulate drug delivery offers significant input in the detection devices and for the categories of disease, which will provide the advancement in drug targeting on a particular disease or disorder without any intrusiveness within the body.
- The metallic NPs such as gold, silver, metal oxides zinc, and iron have exposed great potential regarding biomedical applications, and only due to its greater surface area but also since of their various innate biomedical properties [52].

8.5   DISADVANTAGES OF NANOPARTICLES (NPS)

While the deal with the advantages of nanotechnology, we will also require to think about the disadvantages or negative points of the nanotechnology. It is necessary to understand the disadvantages of particular nanoparticulate drug delivery to avoid failure in the future, while, drug delivery. Currently, nanotechnology cost you lots of many during development and also quite difficult for manufacturing, therefore negative point should be known, is always better than nothing. NPs disadvantages based on their route of administration, nanoparticulate systems, and other aspects as discussed below [53, 54]:

- When NPs administered through the topical route, the loss of high drug content during the application of the particular formulation. The topical route of administration also provides a lack of robust controlled drug delivery.
- NPs when stored over a longer period of time, lots of limitations are born their such as drug removal during storage, the polymorphic transition may happen, particle growth during storage, sometimes formulation contains a high amount of water which is not convenient for administration, limited loading capacity for hydrophilic drugs, etc.
- In addition, small drug particle size and large drug surface area may lead to particles aggregation due to high surface energy, may loss of antibacterial activity in some cases, creates difficulty in physical handling in both dry and liquid preparations, and also results in limited drug loading due to small particle size [55, 56].
- When NPs administered through the parenteral route, the possibility of detection of NPs by reticuloendothelial system cells, they may lead to create a problem with the bioavailability of a particular drug [57].

Nanotechnology provides several advantages as discussed earlier. Despite over disadvantages, nanotechnology extensively used for drug delivery presently, because NPs related disadvantages can be overcome by several ways and they are very less in front of this novel delivery. Several advancements reported in the field of nanotechnology to protect the nanoparticulate system from the aggregation of the particles and storage-related problems.

8.6 APPLICATIONS OF NANOPARTICLES (NPS) IN MICROBIAL PHARMACEUTICS

Accompanying with conventional drug delivery NPs having many practical applications, in the microbial and biomedical field and create large scope in a particular field. Nanotechnology is extensively useful in existing biomedicine, by means of applied in drug delivery. NPs gives not only enhanced drug solubility but also, it provides expressively increased bioavailability by means of encapsulating drug, helps in improved permeation of cell membrane, which shows the good therapeutic level of the drug internally [58]. As broadly growing of this field and knowledge is expressively increased about nanotechnology and multidisciplinary methods employed, the novel targeted drug delivery has been established which gives the drug of selection to obtain the wanted site of action internally. Consequently, it permits the many poorly soluble, impermeable, and highly toxic drugs to reach in a clinical trial as potential therapeutic agents [59]. Presently, lots of applications of NPs are reported related to microbial pharmaceutics. In this segment, applications will be deliberated briefly [16, 60].

8.6.1 NANOPARTICLES (NPS) AGAINST PATHOGENS

Commonly, the drug gives better efficacy while they are delivered through NPs, irrespective of the type of pathogens and drugs. Metallic and non-metallic (organic) both types of NPs provide the good antimicrobial activity of different types of drugs. Targeted NPs show the effect of improved drug concentration intracellularly, up to ten times. Consequently, the drug reaches targeted infected tissue of cellular organelles with or without ligands [61, 62]. Using of NPs, challenging tissues like brain infection, deeper infected cutaneous layers, leishmania, or virus have been targeted easily. Lesser the aggregation in untargeted sites in the body shows lesser in drug toxicity and side effects simultaneously [63]. Lots of antimicrobial agents have decreased their nephron toxicity and hematological difficulties by combined with NPs [64]. As well as the enhancement of drug efficiency, the uses of NPs give more patient-friendly regimens. The amount of dose can be reduced and dosage intervals may increase up to 1–10 days by using NPs such as solid lipid NPs, polymeric micelles, polymeric NPs, and liposomes [65]. These increased intervals maybe because of extended half-life, or extended residence time. Concerning, with residence time, it may also be impacted by the mucoadhesive characteristic of NPs [66, 67]. Loaded drugs half-life and therapeutic effect will be increased by many folds, because of NPs defend them against degradation [68, 69].

8.6.2   SYNERGISTIC EFFECT OF NANOPARTICLE (NP) WITH ANTIMICROBIALS

The conservative chemotherapy technique of acting against infections by using small antibiotic molecule, presently they are very less effective due to its MDR development, nowadays this is a major problem of resistance microbes. By means of a number of first-line antimicrobial agents diminishes, currently last-line antibiotics are widely used to fight against microbial resistance and the number of MDR management options is reduces for patients. Management of last line antibiotic uses is the only option in our hands to combat against the infectious bacteria. A combination of antibiotic molecules into the NPs, maybe metallic NPs (silver, gold, etc.) is a possible method to overcome the problem of microbial resistance by using the synergistic action of the conjugated antimicrobial drug and NPs [70]. Study of AgNPs combined with extensively used antibiotics, such as vancomycin, amoxicillin, and erythromycin, which are active on gram-positive bacteria, which has shown improved antimicrobial activity [71]. Accordingly, several studies demonstrated that the conjugated treatment of small antibiotic drugs with NPs does not always improve in antimicrobial action, it may be drug and NP dependent [72]. Although, these studies, observations, and several findings show its ability for improvement in effectiveness of traditional antimicrobials [16].

8.6.3   NANOPARTICLES (NPS) AGAINST MDR

The antimicrobial resistance of anti-microbial agents is presently increasing, in viral, bacterial, and fungal infections, like candidiasis and

Human Immuno Deficiency Virus. The main use of nanoparticulate drug delivery in such MDR is recovering drug efficiency by avoiding different resistance mechanisms such as efflux pumps, from the wide cell wall, and β-lactamase like candidates. Moreover, increasing the efficiency of drugs, various studies have revealed lower chances of convincing resistance afterward treatment by drug-NP conjugated delivery. Microbial biofilm is a dense bacterial group that sticks to inert or biotic surfaces. Biofilm creation is the main virulence feature for the wide-ranging of pathogens, which cause infections [73]. Biofilm infection is not simply agreeable to current antimicrobial treatments or by the only single magic bullet, methods, since biofilm resistance is a significance of complex biological and physical features with different microbial molecular and genetic features, and they often include multi-species interactions [74]. Biofilm containing microbes promptly produced extracellular polymeric matrix which is difficult to penetrate, therefore increasing the MDR effect. Additionally, they are generally localized at places problematic to reach, thus existing treatments are infrequently successful. Presently, they are accountable for the greatest microbial infections and the good choice, further inhibition, is to eradicate the colony-forming tissue or implant from the body [75]. Few types of drug-NP are capable to cross the bacterial wall and eradicate the biofilm. The multifactorial environment of biofilm growth and drug tolerance enforces the main challenges for the usage of conventional antibacterial and designates the necessity for targeted or conjugated treatments [76, 77]. The nanoparticulate system provides a multifunctional mechanism to attack on bacteria, for this motive, resistance later treats with free drugs are expressively most general than subsequently treatment by drug NPs [65].

8.6.4 NANOPARTICLES (NPS) AS ANTI-INFECTIVE AGENTS

While, an antibiotic is together with anti-microbial NPs, in several cases the effectiveness against sensitive bacteria is high than the alone or free drug. Though, for the management of resistance bacteria, uncoated antimicrobial NP has a greater of similar efficacy than the drug-NP-complex or the free drug. Hence, it is reported that combination drug therapy (NP and drug) against resistant bacteria having improved efficiency. Besides, naked NPs will also favorably inhibit the biofilm and eradicate the bacterial mass and prevent the formation of new bacterial growth. The antimicrobial action of some NPs, especially metallic types of NPs has been studied in both *in vivo* and *in vitro* in different well resistant bacteria. The gold NPs are widely studied, with others like silver [78] and copper NPs [23]. The AgNP gives potent activity against different pathogens, such as antifungal, antiviral, anti-inflammatory, and antibacterial activity [79, 80]. From ancient times, silver is used as medicinal use. As compared with other metal NPs, AgNPs provide greater toxicity towards microorganisms whereas showing less toxicity to human cells [81]. Currently, the most favorable applications have been given in the pharmacological field, like biosensors [82], and for wound healing treatment. In the case of burn treatment, the AgNPs are available in the form of antimicrobial gel preparation meant for the topical burn treatment [83]. The AgNPs are used in the form of NPs gel for the treatment of topical infections, which overcomes the conventional problems of gels and ointments, and simultaneously, nanoparticulate delivery provides greater efficacy [84].

8.6.5 NANOPARTICLES (NPS) FOR GASTRIC INFECTION

Gastric infection is the most common and significant infection in the gastrointestinal tract (GI tract). The conventional drug delivery problem is to come across the GI surface and reach to the bacterial flora. In the last decade of the 19th century, the *H. pylori* bacteria is declared as a carcinogenic agent, therefore it is very necessary to eradicate the infection of *H. pylori* [85]. The mucoadhesive NPs attach with mucin and penetrate well through the cellular junction where infectious bacteria exist. Therefore, the nanoparticulate system increases the residence time in the GI tract and sustained released the drug for an extended period of time to eradicate the microbial biofilm [86]. In addition, target ligand NPs are prepared for the targeting of specific bacteria to avoid unnecessary inflammation in the tract [87]. *H. pylori* bacterial both forms, spiral, and coccoid are eradicated by using a common drug, which acts on both forms of bacteria by nanoparticulate drug delivery. Currently, a combination of several drugs including proton pump inhibitor and antimicrobial agents are administered together to provide good dosing schedule and drug efficiency [88].

8.6.6   NANOPARTICLE (NP) AS VACCINE

It is of most significant to advance strategies for avoiding and treating various infectious diseases. Vaccination is one of the most fruitful medical inventions to avoid and treatment of infectious agents. Recent, vaccines are purified proteins or peptides or recombinant of the analogous pathogen, important to the imitation of the immune response [89]. Nevertheless, they have less immunogenicity and low effect over a long period. This immune response can be enhanced by NPs. Nanotechnology shows the great impact on drug delivery, particularly in the perspective of vaccination. NP facilitated delivery system may competently bind antigen and other adjuvants. Similarly, some materials applied for preparing nanocarriers have an inherent adjuvant effect [90, 91]. Oral vaccination is a preferable route, which may produce systemic and mucosal immune responses, at the same time intestine is the mucosa rich part of the body contains the utmost amount of immune cells. Oral vaccination having the major problem of degradation in the GI tract and fast pass mechanism, therefore oral vaccination requires numerous doses [92]. The present problem can be overcome by coating oral vaccine within NPs. An additional, ideal route for vaccination is via the transdermal route [93]. Skin is the largest tissue of the human body, having lots of immunocompetent cells, capable to stimulate the immune responses. Therefore, cutaneously administered vaccines more efficient than other routes of administration [89]. The penetration capability of NPs depends on properties of the NPs such as the composition of NP, size, and hydric characteristics [94]. Even with their advantages, NPs having some disadvantages, such as early drug release starts, before reaches to the targeted cells, which directly affect the efficacy of the drug [91, 92]. NPs also act as carriers for the adjuvants required in vaccines' long-lasting activity [95]. This technique widely used in several vaccine administrations, this way, vaccines provide long-lasting life, safer, and strong immune response. Antigens also carried by the NPs, to secure from the destructive environment, therefore the vaccine capability is increased by several folds [91].

8.6.7   NANOPARTICLES (NPS) AS THERANOSTIC

Incorporation of the drug within NPs carriers has caused in a novel paradigm for lowering the side effects of chemotherapeutic agents [96]. Theranostic word is made up of with therapeutic and diagnostic approaches whose main motive is the diagnosis, detection, and treatment of the ailments at an initial phase [97]. This is a very novel and promising approach, even for life-threatening diseases. Combining diagnostic and therapeutic agents in several ways, among them combining therapeutic agents with NPs used for imaging purpose and another one is to reserve the agents by encapsulating inside the NPs [98]. Lots of NPs have presented a therapeutic drug and gene delivery transporters as well as diagnostic agent [99]. This nano-sized material gives the opportunity for the detection of several fatal diseases initially and to supervise the therapeutic outcomes of nanostructured drugs used therapeutically. Among many of NPs contains both active ingredient and imaging agent inside a single NP for concurrent disease diagnosis and treatment. Additionally, incorporation of the therapeutic agent along with the diagnostic agent within theranostic NP can be extremely helpful. Though, the distinctive physicochemical features that make NP smart intended for diagnosis and treatment may be as well linked with possible health hazards [100]. Currently, several theranostic systems are used in microbial pharmaceutics, among them NP based theranostic approach less used because of their toxicity and non-existence of understanding of their pharmacokinetics [101].

8.6.8   NANOPARTICLES (NPS) FOR PULMONARY INFECTIONS

Tuberculosis is becoming a recurrent pulmonary infection, due to resistant strains in immunosuppressed patients. Almost 90% of death happens per year, just because of cystic fibrosis. 187 for the solving of this problem, formulation scientists now focus on the various type of NP to increase efficiency and simultaneously to deliver a more comfortable route with modified dosing intervals [61, 102, 103]. The nanoparticulate drug delivery for the treatment of pulmonary tract infection as revealed in Figure 8.3. The general infective bacteria in the pulmonary tract are *Staphylococcus aureus*. The application of NPs marked with antibiotics is very beneficial to treat pulmonary bacterial infection [104]. NPs serve as good carriers for several antibiotics to transport the drug inside the infected pulmonary tract. NPs along with low density (≈ 0.4 g/cm^3) and diameter (>5), presented to be greater capability aerosolized than lesser nonporous particles and prevent phagocytosis [104]. NPs and aerosol properties are essential to avoid exhalation and to confirm that NPs should be placed in the targeted region [105]. Ligand guided NPs are used to target the specific infectious bacteria in the body [106]. The inhaled drug can be exhaled throughout administration; the present problem has been resolved by preparing NPs in inhaled microaggregates. Existing, aggregates stay for an extended period of time in the lungs to gives better antimicrobial action than conventional delivery [107]. The administered NPs need to avoid mucociliary clearance on or after the airways. An approach to enhance the residence time with mucoadhesive NPs, like chitosan NPs, by proper functionalization process. In cooperation, both techniques assist to deepen and upsurge the deposition. Though, so many ex-vivo studies have been performed in the absence of mucous membranes, considering only the NPs properties for sustained release and for interaction with biological barriers [103, 108].

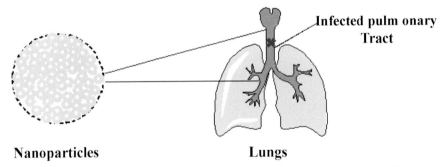

Nanoparticles   **Lungs**

FIGURE 8.3   Targeted nanoparticles in the treatment of the pulmonary tract infection.

8.7   CONCLUSION

NPs exhibit greater potential for therapeutic application in the treatment of various infectious diseases. Existing, literature survey presents that the NPs provide promising treatment in the microbial pharmaceutics. NPs present promising treatment against the microbial infection by targeting the specific infectious microorganism. Additionally, NPs are boosting the outcome of the drugs, even in the condition of the MDR bacterium. Furthermore, it provides lower dosing frequency, enhanced bioavailability, lesser side effects, and a précised administration schedule. The interdisciplinary collaboration is required to develop the nanoparticulate medicine in microbial pharmaceutics. The selection of the NPs depends on drug properties or drug-loaded characteristics and diseased conditions. Likewise, the nanoparticulate drug delivery system should help to develop more effective and more patient convenient dose regiment at the same time they will provide effective drug delivery in the microbial pharmaceutics field. Microbial pharmaceutics is the emerging field in the future, to develop the novel drug delivery systems and NPs based drug delivery is amongst one of the widely used technique. Consequently, NPs based drug administration may consider as a milestone in the field of microbial pharmaceutics.

KEYWORDS

- **dendrimers**
- **drug delivery system**
- **gastric infection**
- **gastrointestinal tract**
- **multidrug resistance**
- **nanoparticle**
- **polymeric nanoparticle**

REFERENCES

1. Tanwar, J., et al., (2014). *Multidrug Resistance: An Emerging Crisis.* Interdisciplinary perspectives on infectious diseases.
2. Rai, M., et al., (2012). Silver nanoparticles: The powerful nano weapon against multidrug-resistant bacteria. *Journal of Applied Microbiology, 112*(5), 841–852.
3. Ranghar, S., et al., (2014). Nanoparticle-based drug delivery systems: Promising approaches against infections. *Brazilian Archives of Biology and Technology, 57*(2), 209–222.
4. Patel, N. R., et al., (2013). Nano preparations to overcome multidrug resistance in cancer. *Advanced Drug Delivery Reviews, 65*(13–14), 1748–1762.
5. Roca, I., et al., (2015). The global threat of antimicrobial resistance: Science for intervention. *New Microbes and New Infections, 6*, 22–29.
6. Zhang, L., et al., (2010). Development of nanoparticles for antimicrobial drug delivery. *Current Medicinal Chemistry, 17*(6), 585–594.
7. Brown, E. D., & Wright, G. D., (2016). Antibacterial drug discovery in the resistance era. *Nature, 529*(7586), 336.

8. Jacob, J. T., et al., (2013). Vital signs: Carbapenem-resistant *Enterobacteriaceae*. *Morbidity and Mortality Weekly Report (MMWR), 62*(9), 165.

9. Laxminarayan, R., & Heymann, D. L., (2012). Challenges of drug resistance in the developing world. *Bmj., 344*, e1567.

10. Davies, J., & Davies, D., (2010). Origins and evolution of antibiotic resistance. *Microbiol. Mol. Biol. Rev., 74*(3), 417–433.

11. He, Q., et al., (2011). A pH-responsive mesoporous silica nanoparticles-based multi-drug delivery system for overcoming multi-drug resistance. *Biomaterials, 32*(30), 7711–7720.

12. Blanco, E., Shen, H., & Ferrari, M., (2015). Principles of nanoparticle design for overcoming biological barriers to drug delivery. *Nature Biotechnology, 33*(9), 941.

13. Laxminarayan, R., et al., (2013). Antibiotic resistance—the need for global solutions. *The Lancet Infectious Diseases, 13*(12), 1057–1098.

14. Soenen, S. J., et al., (2011). Cellular toxicity of inorganic nanoparticles: Common aspects and guidelines for improved nanotoxicity evaluation. *Nano Today, 6*(5), 446–465.

15. Radovic-Moreno, A. F., et al., (2012). Surface charge-switching polymeric nanoparticles for bacterial cell wall-targeted delivery of antibiotics. *ACS Nano, 6*(5), 4279–4287.

16. Kumar, M., Curtis, A., & Hoskins, C., (2018). Application of nanoparticle technologies in the combat against anti-microbial resistance. *Pharmaceutics, 10*(1), 11.

17. Bolla, J. M., et al., (2011). Strategies for bypassing the membrane barrier in multidrug resistant gram-negative bacteria. *FEBS Letters, 585*(11), 1682–1690.

18. Rice, L. B., (2009). The clinical consequences of antimicrobial resistance. *Current Opinion in Microbiology, 12*(5), 476–481.

19. Leaper, D. J., (2006). Silver dressings: Their role in wound management. *International Wound Journal, 3*(4), 282–294.

20. Allaker, R. P., & Ren, G., (2008). Potential impact of nanotechnology on the control of infectious diseases. *Transactions of the Royal Society of Tropical Medicine and Hygiene, 102*(1), 1–2.

21. Li, Y., et al., (2006). Antimicrobial effect of surgical masks coated with nanoparticles. *Journal of Hospital Infection, 62*(1), 58–63.

22. Furno, F., et al., (2004). Silver nanoparticles and polymeric medical devices: A new approach to prevention of infection? *Journal of Antimicrobial Chemotherapy, 54*(6), 1019–1024.

23. Huh, A. J., & Kwon, Y. J., (2011). Nanoantibiotics: A new paradigm for treating infectious diseases using nanomaterials in the antibiotics resistant era. *Journal of Controlled Release, 156*(2), 128–145.

24. Pridgen, E. M., Langer, R., & Farokhzad, O. C., (2007). *Biodegradable, Polymeric Nanoparticle Delivery Systems for Cancer Therapy.*

25. Thierry, B., (2009). Drug nanocarriers and functional nanoparticles: Applications in cancer therapy. *Current Drug Delivery, 6*(4), 391–403.

26. Dong, X., & Mumper, R. J., (2010). Nanomedicinal strategies to treat multidrug-resistant tumors: Current progress. *Nanomedicine, 5*(4), 597–615.

27. Gindy, M. E., & Prudhomme, R. K., (2009). Multifunctional nanoparticles for imaging, delivery and targeting in cancer therapy. *Expert Opinion on Drug Delivery, 6*(8), 865–878.

28. Bhatia, S., (2016). Nanoparticles types, classification, characterization, fabrication methods, and drug delivery applications. In: *Natural Polymer Drug Delivery Systems* (pp. 33–93). Springer.

29. Pal, S. L., et al., (2011). Nanoparticle: An overview of preparation and characterization. *Journal of Applied Pharmaceutical Science, 1*(6), 228–234.

30. Dave, V., et al., (2017). Lipid-polymer hybrid nanoparticles: Development & statistical optimization of norfloxacin for topical drug delivery system. *Bioactive Materials, 2*(4), 269–280.

31. Hasan, S., (2015). A review on nanoparticles: Their synthesis and types. *Research Journal of Recent Sciences, 2277*, 2502. ISSN.

32. Cuenca, A. G., et al., (2006). Emerging implications of nanotechnology on cancer diagnostics and therapeutics. *Cancer, 107*(3), 459–466.

33. Saad, M. Z. H., Jahan, R., & Bagul, U., (2012). Nanopharmaceuticals: A new perspective of drug delivery system. *Asian Journal of Biomedical and Pharmaceutical Sciences, 2*(14), 11.

34. McDevitt, M. R., et al., (2007). Tumor targeting with antibody-functionalized, radio labeled carbon nanotubes. *Journal of Nuclear Medicine, 48*(7), 1180–1189.

35. Wilczewska, A. Z., et al., (2012). Nanoparticles as drug delivery systems. *Pharmacological Reports, 64*(5), 1020–1037.

36. Gao, X., et al., (2004). *In vivo* cancer targeting and imaging with semiconductor quantum dots. *Nature Biotechnology, 22*(8), 969.

37. Moghimi, S. M., Hunter, A. C., & Murray, J. C., (2005). Nanomedicine: Current status and future prospects. *The FASEB Journal, 19*(3), 311–330.

38. Xu, H., et al., (2003). Room-temperature preparation and characterization of poly (ethylene glycol)-coated silica nanoparticles for biomedical applications. *Journal of Biomedical Materials Research Part A: An Official Journal of the Society for Biomaterials, the Japanese Society for Biomaterials, and the Australian Society for Biomaterials and the Korean Society for Biomaterials, 66*(4), 870–879.

39. Freitas, Jr. R. A., (2005). Microbivores: Artificial mechanical phagocytes using digest and discharge protocol. *J. Evol. Technol., 14*, 1–52.

40. Surendiran, A., et al., (2009). Novel applications of nanotechnology in medicine. *Indian Journal of Medical Research, 130*(6).

41. Kalele, S., et al., (2006). Nanoshell particles: Synthesis, properties and applications. *Current Science (00113891), 91*(8).

42. Gao, Z., et al., (2008). Drug-loaded nano/microbubbles for combining ultrasonography and targeted chemotherapy. *Ultrasonics, 48*(4), 260–270.

43. Armstead, A. L., & Li, B., (2011). Nanomedicine as an emerging approach against intracellular pathogens. *International Journal of Nanomedicine, 6*, 3281.

44. Sahoo, S. K., Dilnawaz, F., & Krishnakumar, S., (2008). Nanotechnology in ocular drug delivery. *Drug Discovery Today, 13*(3–4), 144–151.

45. Kost, J., & Langer, R., (2012). Responsive polymeric delivery systems. *Advanced drug Delivery Reviews, 64*, 327–341.

46. Rawat, M., et al., (2006). Nanocarriers: Promising vehicle for bioactive drugs. *Biological and Pharmaceutical Bulletin, 29*(9), 1790–1798.

47. Ahire, E., et al., (2018). Parenteral nanosuspensions: A brief review from solubility enhancement to more novel and specific applications. *Acta Pharmaceutica. Sinica. B, 8*(5), 733–755.

48. De La Escosura-Muñiz, A., et al., (2009). Rapid identification and quantification of tumor cells using an electrocatalytic method based on gold nanoparticles. *Analytical Chemistry, 81*(24), 10268–10274.

49. Sahoo, S. K., Misra, R., & Parveen, S., (2017). Nanoparticles: A boon to drug delivery, therapeutics, diagnostics, and imaging. In: *Nanomedicine in Cancer* (pp. 73–124). Pan Stanford.

50. Smith, A. M., et al., (2006). Multicolor quantum dots for molecular diagnostics of cancer. *Expert Review of Molecular Diagnostics, 6*(2), 231–244.

51. Boisseau, P., & Loubaton, B., (2011). Nanomedicine, nanotechnology in medicine. *Comptes Rendus Physique, 12*(7), 620–636.

52. Wong, K. K., & Liu, X., (2010). Silver nanoparticles—the real "silver bullet" in clinical medicine? *Med. Chem. Comm., 1*(2), 125–131.

53. Parveen, K., Banse, V., & Ledwani, L., (2016). Green synthesis of nanoparticles: Their advantages and disadvantages. In: *AIP Conference Proceedings*. AIP Publishing.

54. Ekambaram, P., Sathali, A. A. H., & Priyanka, K., (2012). Solid lipid nanoparticles: A review. *Sci. Rev. Chem. Commun., 2*(1), 80–102.

55. Beyth, N., et al., (2015). Alternative antimicrobial approach: Nano-antimicrobial materials. *Evidence-Based Complementary and Alternative Medicine, 2015,* p. 246012.

56. Patel, S., et al., (2017). Nanoparticles as a platform for antimicrobial drug delivery. *Advances in Pharmacology and Pharmacy, 5*(3), 31–43.

57. Ghasemiyeh, P., & Mohammadi-Samani, S., (2018). Solid lipid nanoparticles and nanostructured lipid carriers as novel drug delivery systems: Applications, advantages and disadvantages. *Research in Pharmaceutical Sciences, 13*(4), 288.

58. Khadka, P., et al., (2014). Pharmaceutical particle technologies: An approach to improve drug solubility, dissolution, and bioavailability. *Asian Journal of Pharmaceutical Sciences, 9*(6), 304–316.

59. Rodzinski, A., et al., (2016). Targeted and controlled anticancer drug delivery and release with magnetoelectric nanoparticles. *Scientific Reports, 6,* 20867.

60. Ghosh, C. R., & Paria, S., (2011). Core/shell nanoparticles: Classes, properties, synthesis mechanisms, characterization, and applications. *Chemical Reviews, 112*(4), 2373–2433.

61. Kumar, M., et al., (2014). Intranasal delivery of streptomycin sulfate (STRS) loaded solid lipid nanoparticles to brain and blood. *International Journal of Pharmaceutics, 461*(1–2), 223–233.

62. Mahajan, S. D., et al., (2010). Enhancing the delivery of anti retroviral drug "Saquinavir" across the blood brain barrier using nanoparticles. *Current HIV Research, 8*(5), 396–404.

63. Mishra, D., et al., (2014). Glycyrrhizin conjugated chitosan nanoparticles for hepatocyte-targeted delivery of lamivudine. *Journal of Pharmacy and Pharmacology, 66*(8), 1082–1093.

64. Italia, J., et al., (2009). Biodegradable nanoparticles improve oral bioavailability of amphotericin B and show reduced nephrotoxicity compared to intravenous fungizone®. *Pharmaceutical Research, 26*(6), 1324–1331.
65. Gnanadhas, D. P., et al., (2013). Chitosan-dextran sulphate nanocapsule drug delivery system as an effective therapeutic against intraphagosomal pathogen Salmonella. *Journal of Antimicrobial Chemotherapy, 68*(11), 2576–2586.
66. Owais, M., et al., (1995). Chloroquine encapsulated in malaria-infected erythrocyte-specific antibody-bearing liposomes effectively controls chloroquine-resistant Plasmodium berghei infections in mice. *Antimicrobial Agents and Chemotherapy, 39*(1), 180–184.
67. Chhonker, Y. S., et al., (2015). Amphotericin-B entrapped lecithin/chitosan nanoparticles for prolonged ocular application. *International Journal of Biological Macromolecules, 72*, 1451–1458.
68. Xie, S., et al., (2014). Biodegradable nanoparticles for intracellular delivery of antimicrobial agents. *Journal of Controlled Release, 187*, 101–117.
69. Edagwa, B. J., et al., (2014). Long-acting antituberculous therapeutic nanoparticles target macrophage endosomes. *The FASEB Journal, 28*(12), 5071–5082.
70. Deng, H., et al., (2016). Mechanistic study of the synergistic antibacterial activity of combined silver nanoparticles and common antibiotics. *Environmental Science and Technology, 50*(16), 8840–8848.
71. Shahverdi, A. R., et al., (2007). Synthesis and effect of silver nanoparticles on the antibacterial activity of different antibiotics against *Staphylococcus aureus* and *Escherichia coli. Nanomedicine: Nanotechnology, Biology and Medicine, 3*(2), 168–171.
72. Banoee, M., et al., (2010). ZnO nanoparticles enhanced antibacterial activity of ciprofloxacin against *Staphylococcus aureus* and *Escherichia coli. Journal of Biomedical Materials Research Part B: Applied Biomaterials, 93*(2), 557–561.
73. Hobley, L., et al., (2015). Giving structure to the biofilm matrix: An overview of individual strategies and emerging common themes. *FEMS Microbiology Reviews, 39*(5), 649–669.
74. Koo, H., et al., (2017). Targeting microbial bio-films: Current and prospective therapeutic strategies. *Nature Reviews Microbiology, 15*(12), 740.
75. Chen, C. W., et al., (2014). Metal nanobullets for multidrug resistant bacteria and biofilms. *Advanced Drug Delivery Reviews, 78*, 88–104.
76. Høiby, N., et al., (2010). Antibiotic resistance of bacterial biofilms. *International Journal of Antimicrobial Agents, 35*(4), 322–332.
77. Johnson, L. R., (2008). Microcolony and biofilm formation as a survival strategy for bacteria. *Journal of Theoretical Biology, 251*(1), 24–34.
78. Gnanadhas, D. P., et al., (2013). Interaction of silver nanoparticles with serum proteins affects their antimicrobial activity *in vivo. Antimicrobial Agents and Chemotherapy, 57*(10), 4945–4955.
79. Vaidyanathan, R., et al., (2009). *Retracted: Nanosilver: The Burgeoning Therapeutic Molecule and its Green Synthesis.* Elsevier.
80. Panáček, A., et al., (2009). Antifungal activity of silver nanoparticles against *Candida* spp. *Biomaterials, 30*(31), 6333–6340.

81. Zhao, G., & Stevens, S. E., (1998). Multiple parameters for the comprehensive evaluation of the susceptibility of *Escherichia coli* to the silver ion. *Biometals, 11*(1), 27–32.

82. Sun, H., et al., (2009). Nano-silver-modified PQC/DNA biosensor for detecting *E. coli* in environmental water. *Biosensors and Bioelectronics, 24*(5), 1405–1410.

83. Jain, J., et al., (2009). Silver nanoparticles in therapeutics: Development of an antimicrobial gel formulation for topical use. *Molecular Pharmaceutics, 6*(5), 1388–1401.

84. Muangman, P., et al., (2006). Comparison of efficacy of 1% silver sulfadiazine and ActicoatTM for treatment of partial-thickness burn wounds. *Journal-Medical Association of Thailand, 89*(7), 953.

85. Chang, C. H., et al., (2009). Nanoparticles incorporated in pH-sensitive hydrogels as amoxicillin delivery for eradication of *Helicobacter pylori. Biomacromolecules, 11*(1), 133–142.

86. Gonçalves, I. C., et al., (2014). The potential utility of chitosan micro/nanoparticles in the treatment of gastric infection. *Expert Review of Anti-Infective Therapy, 12*(8), 981–992.

87. Lin, Y. H., et al., (2013). Genipin-cross-linked fucose-chitosan/heparin nanoparticles for the eradication of *Helicobacter pylori. Biomaterials, 34*(18), 4466–4479.

88. Ramteke, S., et al., (2009). Amoxicillin, clarithromycin, and omeprazole based targeted nanoparticles for the treatment of *H. pylori. Journal of Drug Targeting, 17*(3), 225–234.

89. Abdul, G. K., et al., (2014). Liposomes as nanovaccine delivery systems. *Current Topics in Medicinal Chemistry, 14*(9), 1194–1208.

90. Bal, S. M., et al., (2010). Efficient induction of immune responses through intradermal vaccination with N-trimethyl chitosan containing antigen formulations. *Journal of Controlled Release, 142*(3), 374–383.

91. Hansen, S., & Lehr, C. M., (2012). Nanoparticles for transcutaneous vaccination. *Microbial. Biotechnology, 5*(2), 156–167.

92. Davitt, C. J., & Lavelle, E. C., (2015). Delivery strategies to enhance oral vaccination against enteric infections. *Advanced Drug Delivery Reviews, 91*, 52–69.

93. Jain, S., et al., (2014). Improved stability and immunological potential of tetanus toxoid containing surface engineered bilosomes following oral administration. *Nanomedicine: Nanotechnology, Biology and Medicine, 10*(2), 431–440.

94. Van, R. E., et al., (2014). Combating infectious diseases; nanotechnology as a platform for rational vaccine design. *Advanced Drug Delivery Reviews, 74*, 28–34.

95. Badiee, A., et al., (2013). Micro/nanoparticle adjuvants for antileishmanial vaccines: Present and future trends. *Vaccine, 31*(5), 735–749.

96. Bilensoy, E., et al., (2008). Safety and efficacy of amphiphilic ß-cyclodextrin nanoparticles for paclitaxel delivery. *International Journal of Pharmaceutics, 347*(1–2), 163–170.

97. Muthu, M. S., et al., (2014). Nanotheranostics-application and further development of nanomedicine strategies for advanced theranostics. *Theranostics, 4*(6), 660.

98. Chen, F., Ehlerding, E. B., & Cai, W., (2014). Theranostic nanoparticles. *Journal of Nuclear Medicine, 55*(12), 1919–1922.

99. Xie, J., et al., (2009). Iron oxide nanoparticle platform for biomedical applications. *Current Medicinal Chemistry, 16*(10), 1278–1294.
100. Ma, X., Zhao, Y., & Liang, X. J., (2011). Theranostic nanoparticles engineered for clinic and pharmaceutics. *Accounts of Chemical Research, 44*(10), 1114–1122.
101. Yoshitomi, T., et al., (2009). pH-sensitive radical-containing-nanoparticle (RNP) for the L-band-EPR imaging of low pH circumstances. *Bioconjugate Chemistry, 20*(9), 1792–1798.
102. Dube, A., et al., (2013). State of the art and future directions in nano medicine for tuberculosis. *Expert Opinion on Drug Delivery, 10*(12), 1725–1734.
103. Mehanna, M. M., Mohyeldin, S. M., & Elgindy, N. A., (2014). Respirable nanocarriers as a promising strategy for anti-tubercular drug delivery. *Journal of Controlled Release, 187*, 183–197.
104. Klinger-Strobel, M., et al., (2015). Aspects of pulmonary drug delivery strategies for infections in cystic fibrosis-where do we stand? *Expert Opinion on Drug Delivery, 12*(8), 1351–1374.
105. Andrade, F., et al., (2013). Nanotechnology and pulmonary delivery to overcome resistance in infectious diseases. *Advanced Drug Delivery Reviews, 65*(13–14), 1816–1827.
106. Unit, W. H. O. M. O. S. A., (2014). *Global Status Report on Alcohol and Health, 2014.* World Health Organization.
107. Sung, J. C., et al., (2009). Formulation and pharmacokinetics of self-assembled rifampicin nanoparticle systems for pulmonary *delivery. Pharmaceutical Research, 26*(8), 1847–1855.
108. Hadinoto, K., & Cheow, W. S., (2014). Nano-antibiotics in chronic lung infection therapy against *Pseudomonas aeruginosa. Colloids and Surfaces B: Biointerfaces, 116*, 772–785.

CHAPTER 9

Spherulites: A Novel Approach in Vesicular Drug Delivery Systems

DEEPAL J. TAANK, NAYANA KALE, SWATI G. TALELE, and
ANIL G. JADHAV

*Sandip Institute of Pharmaceutical Sciences, Nashik, Maharashtra,
India, E-mail: swatitalele77@gmail.com (S. G. Talele)*

ABSTRACT

Spherulites are small, rounded multilamellar lipidic vesicles that can encapsulate biomolecules and may be used as a carrier for drug delivery. Spherulites resemble liposomes but they are obtained by a simpler process (shearing lamellar phases) and have lamellae up to the very center like an onion. Spherulites offer high encapsulation efficiency, cost-effective method of preparation, uniform structure, enhancement of bioavailability, stability, and high reproducibility of the manufacturing process at any industrial-scale over other multilamellar drug delivery systems such as liposomes and niosomes and environment-friendly technology. Spherulites have the ability to incorporate both hydrophilic and lipophilic active molecules without the use of organic solvents. Peptides and protein molecules quickly degraded in the human body, especially in GIT. Therefore, spherulites are used for the delivery of these molecules as their structure allows protecting them from enzymatic degradation. Oral delivery of the anticancer drug is a main concerning pharmacy. The main issues faced are the poor and variable bioavailability of these drugs when administered through the oral route. Spherulites have been used to solubilize highly insoluble anticancer drugs, providing an aqueous pharmaceutical vehicle for these molecules. The present article highlights the formation of spherulites, confirmed by light scattering studies which revealed the birefringent nature of spherulites and multilayered structure. The prepared spherulites were evaluated for drug encapsulation efficacy, drug content, particle size, in vitro drug release, and stability as per ICH guidelines. Thus, this chapter gives emphasis on green and sustainable technology as spherulites are considered as vesicular drug delivery system right from dermatological formulation to vaccines.

9.1   INTRODUCTION

Spherulites are small, rounded multilamellar microvesicles bodies that commonly occur in vitreous igneous rocks (from 0.1–10 μm) [1]. They are frequently visible in specimens of obsidian, pitchstone, and rhyolite as globules to the size of millet seed or rice grain, with duller luster than the surrounding glassy base of the rock, and when examined by a lens they show to have a radiated fibrous structure. Spherulites are commonest in silica-rich glassy rocks [2, 3]. When obsidians are devitrified, and they are traceable, though they may be more or less entirely re-crystallized or silicified. The spherulite center may contain a crystal (e.g., feldspar or quartz) or maybe a cavity. Infrequently spherulites have different color zones, and while most often spherical, they may also be irregular or polygonal in the skeleton. In some rhyolites found in New Zealand, the spherulites send branching cervicorn processes outwards through the surrounding glass of the rock. Axiolites is the name given to long, elliptical, or band-like spherulites [4].

According to their sizes, they are classified as megaspherulite (up to 20 cm in diameter) and lithophysae (up to several cms or more) [5].

9.2   CHARACTERISTICS OF SPHERULITES

- Highly stable and provides protection against enzyme degradation of the incorporated molecule.
- Allows incorporation of both lipophilic and hydrophilic active molecules with high encapsulation yield.
- The encapsulation of fragile molecules like proteins is possible with little stress (pressure, shear, temperature) and without the use of organic solvents.
- Due to their exclusive structural properties and manufacturing process, they have numerous applications like in prolongation,

protection, bioavailability enhancement, administration through a different route, or active substances vectorization.

9.3   PREPARATION OF SPHERULITES

Spherulites are multilamellar vesicles obtained by shearing a lipidic lamellar phase with no aqueous core and whose diameter is comprised between 200 nm and 1 um [6, 7]. Spherulites are constituted by an alterance of lipidic bilayers and aqueous layers and as a result, display an onion-like structure [8]. Initially, phosphatidylcholine (PC), cholesterol, and various cosurfactants constitute their lipidic bilayers as shown in Figure 9.1. A water layer is present in between two bilayers to form a lamellar phase. The shearing of the resultant which is followed by the dispersion of it in aqueous solutions gives the onion-like vesicles called spherulites. Their size is related to their formulation and the shearing's strength and duration.

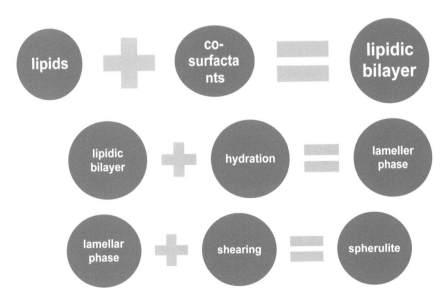

FIGURE 9.1   Spherulites preparation.

9.4   GROWTH OF SPHERULITES

Spherulitic growth is a nonequilibrium process in highly supersaturated solutions, commonly observed in a wide variety of natural and industrial materials [9]. Spherule is formed by the crystallization process which is determined by kinetic parameters that operate under supersaturated conditions and which do not depend on the material type or the crystallographic symmetry. Spherules are composites of littler subunits of fibers or rods which are radially arranged from the development point. The growth of the spherule initiates with unsystematic nucleation on an exterior of seed (a foreign particle, bubble, or a crystal of the same material), and then spherule subunits nucleate on the seed. Just the nuclei that have their c-axes perpendicular to the interface of the solid solution will infiltrate the unstable diffusion layer, because of the quick growth of the c-axial direction, projecting as a finger into the supersaturated zone. This process chooses the spiral growth along the quickest growing direction while the other crystallographic introductions are normally suppressed due to their moderate development rate.

Spherulitic development is regular in nature, yet the elements controlling it were inadequately defined until recently. As shown in Figure 9.2 there are two development types for spherules [10, 11]:

➢ **Type-1:** Polycrystalline spherulite structures become radially outward from a nucleation point, stretching irregularly to keep up a space-filling character. Development along headings from the middle is by growth front nucleation (GFN) [12] after the crystal seed achieves a basic size.

➢ **Type-2:** Polycrystalline spherulite structures develop from a fiber-like structure which branches to shape auxiliary strands framing crystal sheaves, which fan out and near structure a circular development design, and from there on developing in a radial style by GFN. The phase-field theory to describe the poly-crystalline growth is valid for precipitation from supersaturated solutions, including supersaturated aqueous solutions and supercooled melts. For instance, a fluorapatite spherulite has been appeared to frame from a tree fractal type expanding in a fluid gel medium, where the development from a direct seed stretches out to accept a circular shape like Type-2 spherules.

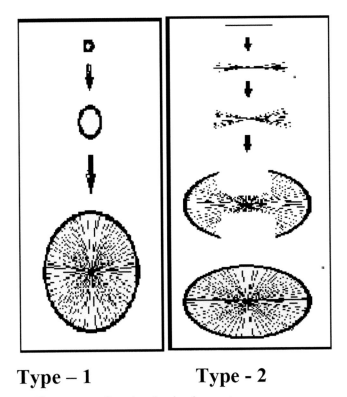

Type – 1   Type - 2

FIGURE 9.2   The two types for spherules development.

9.5   OPTICAL PROPERTIES

Because of the crystallinity present inside the spherulite structure and their high degree of the order, they display birefringence [13]. By introducing the polymer spherulite between crossed polarizers, different colors can be seen.

9.5.1   BIREFRINGENT OF SPHERULITES AND MALTESE CROSS

Spherulites are best seen in a microscope when kept between crossed polarizers. The field seems to be dark if no birefringent taster is set between the polarizers. Anyway, since spherulites are optically anisotropic and crystalline in nature, they translate the plane-polarized light transmitted by the first polarizer to elliptically polarized light. This presently has a partial parallel to the second polarizer (analyzer) which is permitted through. Spherulite along these lines seems bright, with a distinctive black "Maltese cross" [13].

The birefringence of the spherulite is defined as:

$$\Delta n = n// - n_{\perp}$$

where: n = refractive index parallel to the chain axis; n = refractive index perpendicular to the chain axis.

9.5.2   RINGED OR BANDED SPHERULITES [14, 15]

Spherulites demonstrate an extra substituting dark and bright concentric ring structure that is credited to the expected twisting of the radial crystallite strips as they develop from the spherulitic focus. This is on the grounds that as they develop radially the crystalline lamellae likewise turn around their radial development axis. The necessity of space-filling implies that the twirl must be synchronized [16, 17]. Thus, at a given distance from the core, all strips must be twisted by the equal angle to stay parallel to one another. This ultimately leads to the formation of the concentric circles (actual spheres in 3d). The spherulite with such a composite pattern is so-called the banded (ringed) spherulite.

9.6   EVALUATION

9.6.1   DRUG ENTRAPMENT EFFICIENCY OF LIPOSOMES [18, 19]

DEE is determined by the centrifugation method. Samples (1 ml) of spherulite dispersion were centrifuged on a laboratory centrifuge at a rate of 3500 rpm for a time period of 90 min. The supernatants were then removed with caution in order to separate out non-entrapped drug and absorbance were recorded. % Drug entrapment was calculated by using the formula:

9.6.2   SPHERULITES SIZE AND ZETA POTENTIAL DETERMINATION

The size and zeta potential of the spherulites were determined by dynamic light scattering using a Malvern size analyzer. Spherulites dispersion was prepared and suitably diluted with distilled water and the mean hydrodynamic diameter with poly-dispersity index (PDI) was computed [18].

9.6.3   TRANSMISSION ELECTRON MICROSCOPY (TEM) [18]

Spherulites morphology and internal structure can be monitored by transmission electron microscopy (TEM) operated at 200 kV with magnification power up to 75,000. The sample is prepared by adding a drop of the formulation on the grid and allowed to dry. The prepared grid was mounted on a single tilt holder and inserted into the TEM instrument.

9.6.4   IN VITRO DRUG RELEASE FOR SPHERULITES [18]

In vitro drug release is studied using phosphate buffer saline (PBS) using the dialysis membrane. Dialysis bag is immersed in PBS overnight in order to activate the bag which is then followed by rinsing it with distilled water. The bag is then cut into three pieces and each is then filled with spherulitic solutions. The bags are clipped using closure clips at both ends. These are then suspended in 100 ml of PBS of pH 7.4 in a beaker and are stirred at 100 RPM using magnetic stirrer with the temperature maintained at 37°C. Aliquots are withdrawn at previously determined intervals and drug amount in each is determined by UV spectrophotometric technique.

9.6.5   IN VITRO AND IN VIVO STABILITY STUDIES OF SPHERULITES

A major prerequisite in the use of lipidic vesicles as drug carriers in vivo is that they must both circulate and retain drugs for periods of time sufficient for effective access to, and interaction with the target, usually in the blood, lining the capillaries and, in certain cases, cells in extravascular areas. However, a considerable drawback of the use in clinical practice is the poor stability of the liposomes in the blood circulation. Peptides and proteins are easily hydrolyzed by a large number of proteases and peptidases present in the serum or gastrointestinal tract (GI tract) depending on the administration route [20, 21]. Thus, in order to promote oral administration, systems protecting the drugs from peptidases are required. The few interesting results obtained with the liposomes can be attributed to the lack of liposomal stability in the environment of the GI tract [22]. The bile salts are able to disrupt and solubilize the lipidic vesicles membrane [23]. The entrapped drugs were no longer protected from peptidases present in the GI tract.

O. Diatet, have investigated the encapsulation of protein A in a new multilamellar vesicle (spherulites) composed of PC, cholesterol, and polyoxyethylene alcohol (Cl_2H_25 (OCH_2CH_2)40H) as a neutral detergent in relation with the preparation method [24]. Spherulites entrapping protein A were prepared by shearing a phospholipidic lyotropic lamellar phase. It was observed that shearing a lyotropic lamellar phase leads to the formation of a very interesting structure similar to liposomes. By diluting these preparations, it separates out single spherulites without any change in their internal multilamellar. This process permitted to obtain a size of about 300 nm depending upon the shear rate [25]. To keep a check on the intact transportation of the spherulite into the general circulation, it was labeled with radioactive substance indium and the radioactivity in the blood was counted. Free indium can penetrate the intestinal wall but does not enter the blood circulation [26]. If the spherulites were disrupted, indium will be rejected and radioactivity would not be detectable in the blood and thus the use of radioactive indium helped us to differentiate whether the spherulites are broken down or not in the epithelium. Various experiments seem to show that the spherulites formulation proposed presents interesting resistance in gastric and intestinal fluid and that some of the spherulites penetrating the epithelial wall are not lysed.

9.7 STERILIZATION

Spherulites are new promising multilamellar vesicles and designed to contain and deliver drugs by local or parenteral injection, they are supposed to be sterile. Terminal heat sterilization, gamma-irradiation, UV irradiation, ethylene oxide treatment, high-pressure sterilization, and sterilizing filtration have been reported [27]. Antoine Richard et al. [28] discussed that sterilizing filtration is a kind of pharmaceutical operation that is aimed at eliminating particles and microorganisms from within a liquid by use of porous filter in order to obtain a sterilized suspension or solution using Sterilizing-grade filters. Several types of filters are commercially available which is given as follows:

1. Hydrophobic PTFE membranes to filter organic solutions.
2. Hydrophilic PTFE membranes to filter aqueous or aqueous-organic solutions.
3. Hydrophilic PVDF membranes for the filtration of aqueous solutions where the proteic adsorption is minimized (Figure 9.3).

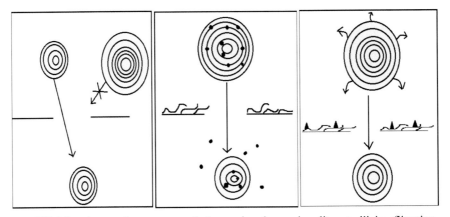

FIGURE 9.3 Assumed occurrences that can arise along spherulites sterilizing filtration.
(a) Allowance of only small-sized spherulites.
(b) Escape of a formerly encapsulated compound due to the disruption of external bilayers.
(c) Damage of the MRP and/or chemical functionality due to the disruption of external bilayers.

9.8 APPLICATION

9.8.1 IN DERMATOLOGY

Spherulites are utilized for two primary effects in dermatology [29]. A reservoir effect is acquired utilizing cationic added substances among the parts of the amphiphilic membrane. This positively charged vesicles interface with the contrarily charged protein of the skin or of the hair so as to expand the contact time of the active ingredient with a targeted site of administration. Another effect is connected to the capacity of the Spherulites to co-encapsulate permeation enhancers together with the active substance. The idea of the amphiphiles establishing the vesicle and the addition of explicit enhancers permit to alter the permeation ability of the active substance. An antiviral medication dependent on this idea is a work in progress.

9.8.2   IN VACCINE

Antigen encapsulation in spherulites prompts high titers of antibodies in mice vaccinated by parenteral course. Spherulites turn out to be a potent transport system for systemic responses. In addition, Spherulites can inspire a reasonable reaction with potentiation of IgG2a antibodies as per higher IFN-γ potentiation by splenocytes [30].

Intranasal administration of antigen contained within spherulites induces definite IgA antibody response not only in the lungs but also in other distant mucous spots, i.e., in vagina and in intestine. In addition, intranasal administration of spherulites produced both IgG and IgA antibodies in serum demonstrating that spherulites can likewise be utilized for promoting a systemic response by a non-intrusive way.

9.8.3   IN PEPTIDE AND PROTEINS DELIVERY

F. W. Billmeyer elaborated peptide and proteins are delicate particles that immediately degrade in the human body, particularly in the gastrointestinal tract [31]. Huge exertions are made to discover a non-intrusive technique for the conveyance of these powerful medications. Spherulites are all around intended for the conveyance of these molecules as their structure permit to shield them from enzymatic degradation [32, 33].

9.8.4   ANTICANCER DRUGS DELIVERY

The development of oral drug delivery of the anticancer drug is one of the biggest challenges in pharmacy today [18]. Poor and variable bioavailability of oral medication can be overwhelmed by spherulites an innovative system to solubilize profoundly insoluble anticancer medications like camptothecin subordinates, in this way giving an aqueous vehicle to these molecules.

9.8.5 SHIGA TOXIN B-SUBUNIT

Intracellular conveyance methodology has been grown as of late utilizing the non-harmful receptor-restricting STxB (Shiga poison B-subunit), which is delivered by *Shigella dysenteriae* and enterohemorrhagic strains of Escherichia coli [34]. STxB (Shiga poison B-subunit) is known to tie the glycosphingolipid Gb3 (globotriaosylceramide), which is overexpressed by different human tumors. After Gb3 official, the poison enters the cytoplasm by means of the retrograde course, bypassing the debasing condition of the late endosomes/lysosomes. Anthony Bouter et al. examined the functionalization of spherulites by STxB and the intracellular dealing. 0STxB-spherulites (ST×B-functionalized spherulites) are disguised into HeLa cells in a receptor-subordinate way. STxB disturbed plasma film, prompting lipid and substance focusing on the established endocytic pathway.

9.8.6 ENZYME MICROENCAPSULATION [35]

Bernheim-Grosswasser et al. elaborated encapsulation of alkaline phosphatase effectively into spherulites and after the destruction of vesicles, the whole activity could be get recovered and utilized for industrial and medical applications. Evaluation parameters revealed that encapsulation efficiency with the size of spherulites and mechanical strength can be preserved efficiently.

9.8.7 GASTROINTESTINAL DRUG DELIVERY [36]

O Freund discussed spherulites preferred over liposomes. Protein and peptide when administered as liposomes, bile salt causes disruption of vesicle membrane, and drugs may not be protected because of peptidase. In contrast, spherulites successfully prepared by the process of shearing as deliberated above and biodistribution study was carried out [125]. I-Protein

A utilized to conduct a biodistribution study of spherulites by administering intravenously. After intravenous administration scientists observed promising results, immediately radioactivity gets cleared from circulation shows more accumulation in liver and spleen. This indicates spherulites could be an effective approach for gastrointestinal medicaments. Valberg et al. [37] investigated indium aid in the detection of spherulites as lysed or nonlysed [111]. In NTA, entrapped spherulites administered orally and observed noteworthy radioactivity in blood.

9.9   CONCLUSION

Due to their exclusive structural properties and manufacturing process, they have numerous applications like in prolongation, protection, bioavailability enhancement, administration through a different route, or active substances vectorization. Spherulites offer high stability and provide protection against enzyme degradation of the incorporated molecule. Spherulites were stable in an acidic (pH 3) and basic buffer (ph 10) when they were incubated. In prospect, ample scope will be available for spherulites as it permits the incorporation of both lipophilic and hydrophilic active molecules with high encapsulation yield. The encapsulation of fragile molecules like proteins is possible with little stress (pressure, shear, temperature) and without the use of organic solvents. For pharmaceutical applications, so as to reduce toxicological issues, spherulites are manufactured starting from already ratified components. The technology is already being used in the field of cosmetics and veterinary.

KEYWORDS

- **growth front nucleation**
- **liposome**
- **microvesicles**
- **multilamellar**
- **poly-dispersity index**
- **spherulite**

REFERENCES

1. Baird, B., (1990). *Stone Spheres* (p. 24). The Edinburgh Geologist, Spring.
2. Heinrich, P. V., (2007). *Megaspherulites* (Vol. 38, No. 7, pp. 8–12). PDF Version, 1.4 MB Back Bender's Gazette.
3. Smith, R. K., Tremallo, R. L., & Lofgren, G. E., (2000). *Megaspherulite Growth: Far From Equilibrium Crystallization.* GeoCanada—The Millennium Geoscience Summit, Canadian Society of Exploration Geophysicists Annual Meeting.
4. Smith, R. K., Tremallo, R. L., & Lofgren, G. E., (2001). Growth of megaspherulites in a *Rhyolitic vitrophyre. American Mineralogist, 86*(5–6), 589–600.
5. Rodríguez, E. A., (2002). *The Fantastic Balls in el Cerro Piedras Bola (Jalisco)* (p. 305). México Desconocido.
6. Diat, O., & Roux, D., (1993). Preparation of monodisperse multilayer vesicles of controlled size and high encapsulation ratio. *J. Phys. II France, 3*, 9–14.
7. Gulik-Krzywicki, T., Dedieu, J. C., Roux, D., Degert, C., & Laversanne, R., (1996). *Freeze-Fracture Electron Microscopy of Sheared Lamellar Phase* (Vol. 12, pp. 4668–4671). Langmuir.
8. Diat, O., Roux, D., & Nallet, F., (1993). Effect of shear on a lyotropic lamellar phase. *J. Phys. II France, 3*, 1427–1452.
9. Magill, J. H., (2001). *J. Mater. Sci., 36*, 3143.
10. Padden, F. J., & Keith, H. D., (1965). *J. Appl. Phys., 36*, 2987.
11. Norton, D. R., & Keller, A., (1985). *Polymer, 26*, 704.
12. Gránásy, L., Pusztai, T., Börzsönyi, T., Warren, J. A., & Douglas, J. F., (2004). *Nature Mater, 3*, 635.
13. David, I. B., (2002). *An Introduction to Polymer Physics* (pp. 133–136). Cambridge University Press. ISBN 0-521-63721-X.
14. Keith, H. D., & Padden, F. J., (1984). Twisting orientation and the role of transient states in polymer crystallization. *Polymer, 25*, 28–42. (Google Scholar, Cross Ref.).
15. Lotz, B., & Cheng, S. Z. D., (2005). A critical assessment of unbalanced surface stresses as the mechanical origin of twisting and scrolling of polymer crystals. *Polymer, 46*, 577–610. (Google Scholar, Cross Ref.).
16. Shtukenberg, A. G., Punin, Y. O., Gujral, A., & Kahr, B., (2014). Growth actuated bending and twisting of single crystals. *Angew. Chem. Int. Ed., 53*, 672–699. (Google Scholar, Cross Ref., Pub. Med.).
17. Crist, B., & Schultz, J. M., (2016). Polymer spherulites: A critical review. *Prog. Polym. Sci., 56*, 1–63. (Google Scholar, Cross Ref.).
18. Dhande, R., Tyagi, A., Sharma, R. K., & Thakkar, H., (2017). Biodistribution study of 99mTc-gemcitabine-loaded spherulites in Sprague—Dawley rats by gamma scintigraphy to investigate its lung targeting potential. *Journal of Microencapsulation, 34*(7), 623–634.
19. Immordino, M. L., Dosio, F., & Cattel, L., (2006). Stealth liposomes: Review of the basic science, rationale, and clinical applications, existing and potential. *Int. J. Nanomed., 1*(3), 297–315.
20. Kararli, T. T., (1989). *Crit. Rev. Ther. Drug Carrier Syst., 6*, 39–86.
21. Senior, J., & Gregoriadis, G., (1982). *Life Sciences, 30*, 2123–2136.

22. Richards, M. H., & Gardner, C. R., (1978). *Biochim. Biophys. Acta, 543*, 508–522.
23. Nagata, M., Yotsuyanagi, T., & Ikeda, K., (1988). *Chem. Pharm. Bull., 36*, 1508–1513.
24. Diat, O., Roux, D., & Nallet, F., (1993). *J. Phys. II France, 3*, 1427–1452.
25. Diat, O., & Roux, D., (1993). *J. Phys. II France, 3*, 14.
26. Valberg, L. S., Flanagan, P. R., Haist, J., Frei, J. V., & Chamberlain, M. J., (1993). *Clin. Invest. Med., 4*, 103–108.
27. Zuidam, N. J., Versluis, C., Vernooy, E. A. A. M., & Crommelin, D. J. A., (1996). Gamma-irradiation of liposomes composed of saturated phospholipids. Effect of bilayer composition, size, concentration, and absorbed dose on chemical degradation and physical destabilization of liposomes. *Biochim. Biophys. Acta-Biomembr., 1280*, 135–148.
28. Antoine, R., Julie, D., & Line, B. B., (2006). Effects of sterilizing-grade filters on the physicochemical properties of onion-like vesicles. *International Journal of Pharmaceutics, 312*, 144–150.
29. Crauste-Manciet, S., Khawand, K., & Mignet, N., (2015). Spherulites: Onion-like vesicles as nanomedicines. *Therapeutic Delivery, 6*(12), 1377–1385.
30. Choi, Y., Jeon, Y. H., Kang, J. H., Chung, J. K., Schmidt, M., & Kim, A. C., (2007). MIDGE/hNIS vaccination generates antigen-associated CD8+IFN-γ+ T cells and enhances protective antitumor immunity. *Int. J. Cancer, 120*, 1942–1950.
31. Billmeyer, F. W., (1984). *Textbook of Polymer Science* (3rd edn., pp. 273–281). John Wiley & Sons, NY.
32. Gopferich, A., & Langer, R., (1993). *J. Polymer Sci., 31*, 2445–2458.
33. Achim, G., Ruxandra, G., Yoshiharu, M., Lisa, S., Maria, J. A., Yashuhiko, T., & Robert, L. (1994). *Drug Delivery from Bioerodible Polymers Systemic and Intravenous Administration* (p. 567). Chapter 15, ACS symposium series.
34. Lingwood, C. A., Law, H., Richardson, S., Petric, M., Brunton, J. L., De Grandis, S., & Karmali, M., (1987). Glycolipid binding of purified and recombinant *Escherichia coli* produced verotoxin *in vitro*. *J. Biol. Chem., 262*, 8834–8839.
35. Bernheim-Grosswasser, A., Ugazio, S., Gauffre, F., Viratelle, O., Mahy, P., & Roux, D., (2000). Spherulites: A new vesicular system with promising applications. An example: Enzyme microencapsulation. *The Journal of Chemical Physics, 112*(7), 3424–3430.
36. Freund, O., (2001). Biodistribution and gastrointestinal drug delivery of new lipidic multilamellar vesicles. *Drug Delivery, 8*, 239–244.
37. Valberg, L. S., Flanagan, P. R., Haist, J., Frei, J. V., & Chamberlin, M. J., (1981). Gastrointestinal metabolism of gallium and indium. *Clin. Invest. Med., 4*, 103–108.

CHAPTER 10

Role of Omega-3 Fatty Acids in Different Neurodegenerative Disorders

EKNATH D. AHIRE, KHEMCHAND R. SURANA,
CHANDRASHEKHAR D. PATIL, HENNA S. SHAH,
GANESH B. SONWANE, and SWATI G. TALELE

Sandip Institute of Pharmaceutical Sciences, Nashik, Maharashtra, India, E-mail: swatitalele77@gmail.com (S. G. Talele)

ABSTRACT

Omega-3 fatty acids are related to healthy aging all over life. A moment ago, from fish, source omega-3 fatty acids DHA and EPA have been related with fetal development in AD and cognition related problems. Nevertheless, since the human body does not effectively produce few omega-3 fatty acids, while obtaining from marine sources. The brain is extremely augmented with lipids bilayers. Therefore, it is accountable to undertaken that the combination of different fatty acids within brain ha importance for brain functioning, counting neuropsychiatric and cognition development. To know the special effects form fatty acids ingestion, we need to depend on the mutual evaluation of experimental studies, observations, and interventional studies. The brain is enriched with lipid bilayers of omega-3 fatty acid constituents like DHA and EPA may have lots of effects on the brain. The potential function of omega-3 fatty acid, DHA, and EPA in the inhibition of cognition degeneration including AD has involved key importance from the last few decades. Additionally, AD has delivered the utmost positive sign to backing the situation that inflammation gives to a neurodegenerative disorder. The main etiology of the AD is unidentified, but genetic and ecological parameters are supposed, like family history, poor lifestyle and poor diet, the existence of any chronic disease like cardiovascular disease, increased age, and diabetes. It is assumed that primarily, management or prevention of inflammation may delay the indications of AD. Normally, changes the brain physiology with increased aging, comprising reduction of elongated chain omega-3 fatty acids as well as the brain of AD patients have reduced the DHA presence. Provided supplementary DHA can decrease the indications of inflammations. The health-related benefits of omega-3 fatty acids are really appreciable. Presently, there is no evidence of maintenance or confirmation of that modifiable risk parameters like herbal medicines, nutritional supplements, modified diet, etc. are associated with reduced the risk factors of AD or the cognition failure. Even though, multiple studies are showing that lifestyle and diet-related parameters are the main risk factors for AD and cognition behavior. This chapter basically focused on the different functions of omega-3 fatty acids mainly DHA and EPA in different neurodegenerative diseases and disorders.

10.1 INTRODUCTION

10.1.1 NEURODEGENERATIVE DISORDER

Neurodegenerative disorders, followed by progressive and chronic, are described by symmetric and selective damage of neurons in cognitive or sensory, motor systems. It is an incident that happens in the CNS via the feature associated with damage of neuronal composition and function [1]. Neurodegeneration is experienced after viral infection and mostly in neurodegenerative diseases; usually experienced in geriatric patients like Alzheimer's disease (AD), Parkinson's disease (PD), etc.—the contributory mediators of neurodegeneration are not yet perfectly identified [2, 3]. On the other hand, recent data have recognized that the inflammation phenomenon being stringently connected with different neurodegenerative ways, which are related with cognition, AD, an outcome of neurodegenerative disorder. Existing data suggest that the functions of neuron inflammation in neurodegeneration should be totally revealed, while pro-inflammatory mediators, which are the contributory outcome of neuron inflammation [3, 4]. In neurodegenerative disorder, neural injury is the main contributor to various diseases such as cognitive dysfunction, AD, PD, etc. For most neurodegenerative disorders, inflammation may initiate moreover from unidentified stresses related to sporadic diseases

[5]. The most translational investigation into neurodegenerative disorders is dedicated to developing medicines, which can prevent neuronal dysfunction and early death in the disease state. Well-understanding of the molecular and cellular processes of neuronal damage has to be identified for the specific target [6]. Different approaches are there for the blocking of the apoptotic triggers, among them supplementation related to omega-3 fatty acids is the one of stress-free and greatest. Even though this is an exhilarating era in the branch of neurodegenerative disorders, along with molecular and genetic approaches quickly preceding our thoughtfulness of disorder pathogenesis, because there are no other operative treatments for any of the disorders explained earlier [7]. Consequently, there is a requirement of methods development, which will reduce the risk or preventing for neurodegenerative disorders. The mechanism through which dietary restraint gives neurons may encompass a reduced level of mitochondrial oxygen radical or a minor metabolic tension response, in which neurons replied to the tension of abridged energy availability by enhancing the production of tension proteins and some neuronal factors [6, 8, 9].

10.1.2   COGNITIVE DISORDER

The cognitive disorder is a major disorder in the public due to a lack of quality of lifestyle in day-to-day life. This disorder observed in aged populations associated with AD. Globally 35.6 population suffering from Dementia and among that 88% of the population is associated with partially Dementia also known as cognition disorder. In several recent studies, it is described that the deficiency of omega-3 polyunsaturated fatty acid (PUFA) effects on cognition. Clinical studies are done to show the proof of therapeutically and clinically value of omega-3 PUFA in neurodegenerative diseases such as AD. This disorder is found in the people who are aged or elder and there is a requirement of treatment at an initial stage of the disorder. Therefore, the effectiveness, safety, and benefits of omega-3 polyunsaturated fatty acids in cognitive disorders are studied by using recent studies [10]. The purpose of the study to improve cognitive performance and mood-related disorders globally. This study is one of the comparative studies of mood-related disorder especially cognition in the community. The cognitive disorder is a major bipolar disorder because of poor quality of life and due to functional impairment. Previous study there was mentioned that the demonstration of the study is compared with controls [11]. Cognitive disorder creates a problem related to functional impairment and changes in mood disorder. Therefore, the cognitive disorder is a major goal for the study of this topic [12]. Bipolar disorder impacts a 1–3.8% of the all-over population. In the case of the foremost depressive disorder, it ranges from 4.7% to 10.7% of the whole population. One of the previous studies shows the cognitive related problems in Brazil, they study these mood-related disorders in patient and showed a major impact of mood-related disorders. They also compile the data related to the population suffering from bipolar disorder and cognitive impairment. In another study, they also evaluate the different stages of the disorder found in the subjective study in the early stage and long-term stage. Late-stage illness cognitive disorder shows greater impairment cognitive performance as compared to the control group [13].

Cognitive weakening is a significant parameter of psychiatric disorders such as bipolar conditions, cognition, depression, and schizophrenia [14, 15]. Antipsychotic medicaments utilized to treat the neurodegenerative disorders may provide the greatest improvement in cognitive behavior [16]. While, the efficiency of present medicines for the controlling of cognitive indications is quite a bit controversial [17]. Therefore new medications required to the management of the cognitive impairment and some studies were done with regarding to this [18, 19]. Age-related parameters of cognitive function are detected in humans [20]. Actually, aging is the utmost significant risk parameter for the impairment of cognitive related disorders [21]. Cognition disorder is a major problem presently observed in elderly patients and the treatments available for these disorders are ineffective due to some interference. In the market there are several products are available related to recognition enhancement in children's but very few or negligible products are there in the market for the elderly as well as adult patients. Therefore, it is wide scope for the formulation scientist to develop such products which will be modifying the recognition within the elderly as well as adult patients [22].

10.1.3   ALZHEIMER'S DISEASE

Alzheimer's disease (AD) is one of the unrecoverable degenerative diseases of brains that directly affect brain function. Alzheimer's suffered peoples face the many basic problems like recognition meanings of simple words and lost their ability to memories past crusted events. Patients associated with AD in early-stage gradually observed the loss of semantic memory. The AD shows more impairment with semantic fluency than phonemic fluency and semantic fluency may get fast degraded than phonemic fluency. Word association study shows that the AD-associated people lose their own control with simple wording and their meanings can't recognize properly. Rate of AD increases as directly increasing the impairment of word confusion or semantic association [23]. AD creates large numbers of the hypothesis but unfortunately, no anyone can found its proper pathogenesis, and the cause of rapid neurodegeneration results in the formation of senile plaques. AB oligomers accountable for synaptic impairment betoken that the oligomers are only, responsible for the loss of signals that affecting brain functionality. Previous studies reported that cause of the AD by several other mediators iron overload, cholesterol level, innate immunity, oxygen-free radicals, AB oligomers that triggers the action of microglial cells. AD is associated with multiple types of factors neuroinflammation, protein alteration, oxidative stress, immune deregulation, neuronal impairment glial communications that are resulting in neurological degeneration [24]. The unwell myelinated elongated axon hippocampal, as well as cortical neuron, demands high energy and that affects the degeneration of AD. Energy consumption is more in the case of cognitive disorder and it further goes too developed in Alzheimer. Alpha-hydroxybutyrate and Beta-hydroxybutyrate deliver energy to the brain in case glucose uptake is not possible, providing the regular energy to the brain it may lead to delay the action [25].

AD is a very disturbing ailment for which there are inadequate treatment opportunities and not perfectly treat. Recognition loss is an initial symptom of the AD, which is advanced, and provides the incapability of the patient to take carefulness for them or himself and ultimately they cause death [26, 27]. Presently, many studies are conducted on the management of AD in a new formulation development era. The docosahexaenoic acid (DHA) exists in great quantity in the neuronal lipid membrane, wherever it is participated in the appropriate role of the nervous systems to regulate the AD [28]. The neurodegenerative disorder sowing the AD brain and neurons and normal brain and neuron, as revealed in Figure 10.1.

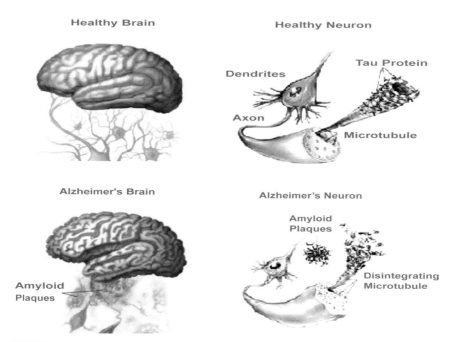

FIGURE 10.1   Normal (healthy brain) and Alzheimer's brain, and healthy neuron and Alzheimer's neuron.

10.2   FUNCTIONS OF OMEGA-3 ACIDS IN NEURODEGENERATIVE DISORDERS

The omega-3 fatty acid has been associated in life, which provides strong healthy aging all over the lifecycle. Recently, DHA and eicosapentaenoic acid (EPA) constituents of omega-3 fatty acid obtained from the marine (fishes) associated source, which shows significant development in different neurodegenerative disorders [29, 30]. DHA and EPA are incorporated in various body parts comprising cell membranes and also playing an essential role in viscosity of the cell different membranes [31, 32]. DHA is an important ingredient of cell membranes and mostly obtained in eye retina and brain cell membranes [33]. It may be very challenging to acquire sufficient consumption of DHA and EPA via food alone, even they are obtained from the water plants like marine animals and algae.

Another one shorter chain containing constituent of omega-3 fatty acid, alpha-linolenic acid (ALA), is a noticeable element of our food, as it is found in numerous land plants which are commonly eaten by a human, however, it does not offers health aids observed with DHA and EPA [34]. While, it can be possible for the human body to transform ALA to DHA and EPA by using desaturase and elongase enzymes, but earlier reported research shows that only little quantity of enzymes is synthesized in the body for this practice, which is not sufficient to conversion [35]. Involuntary, weight loss is the foremost problem with numerous patients with AD may face, the supplement containing DHA and EPA gives patients weight gain significantly in this case. According to recent studies, DHA and EPA may provide different functions with cognitive functioning, weight management, and prevention of mild type of AD. While, results obtained from the different studies concerning the disease caused, due to AD-like cognitive dysfunctioning is mange promisingly by using the omega-3 fatty acids [36]. Patients having AD shown to be an insufficiency of DHA and EPA, by treatment with supplements of DHA+EPA do not only recover the deficiency, but also improved the cognition properties in patients with minor AD [34].

In aged populations, cognitive deficiency accompanying with dementias and AD has developed a considerable problem. It's presently estimated that around 35.6 million patients are surviving with dementia all over the world [37]. Although, these disturbing statistics, it is predicted that postponing of disease treatment develops many patients with the same. AD is primarily categorized by amnesia executive impairment and developments to global shortfalls that eventually lead to total incompetence [38]. Recent studies designate that the higher ingestion of omega-3 fatty acids is accompanying with a lesser rate of cognitive failure and maybe defensive in contradiction of the onset of cognitive deficiency and neurological indisposition [39]. The omega-3 fatty acid management was accompanied by a small, but beneficial, important for instant attention, recall, and processing velocity, persons with mild AD [40]. Omega-3 fatty acids have significant functions in the management and prevention of different human neurodegenerative diseases. Their protective properties on the brain late in life in contradiction of disorders like AD are unidentified but they are positively praiseworthy of study. Preclinical and epidemiological studies designate that the ingesting of elongated chain omega-3 fatty acids may show the cognition weakening and avoid the advancement of mental health ailments such as AD. The correlation between omega-3 fatty acids and mental health disorders has been shown with a lesser extent of omega-3 fatty acids with the plasma of the patients suffering from the neurodegenerative diseases as associated with healthy people [41–43].

10.3   SOURCES OF OMEGA-3 FATTY ACID

The functions of omega-3 fatty acids are most focused to treat or to prevent different neurodegenerative disorders. The outcomes of omega-3 fatty acids with cognitive function in aging, dementia, and neurological disorder are appreciable. Quadragenerian People (above 40 years) suffer from several physiological dysfunctions that have been allocated omega-3 fatty acid, have a major role in the movement of calcium ions in the cells, due to this calcium ions contraction & relaxation of the smooth muscles are takes place [44]. The omega-3 fatty acid contains different chemical constituents. It is mainly composed of the main three different chemical constituents, namely, a) alpha-linolenic acid (ALA), b) eicosapentaenoic acid (EPA), c) DHA. Some selective food is a good source of omega-3 fatty acids, which should be provided to the body as a supplementary source to fulfill the requirement of omega-3 fatty acids [45]. DHA, as well as EPA, are polyunsaturated fatty acids, which are found mostly in aquatic sources as well as in algae. Together DHA and EPA are both significant fatty acids that come into the body via the ingestion of fortification and marine products. Many studies disclose that these fatty acids give a significant function for maintaining a healthy body [46]. Increasing the requirement of DHA and EPA enclosing fish oil sources is a knocking burden on fish species. Many fisheries deliver fishes for the human ingestions, therefore fishes for the purpose of preparation in medicine make shortage. Therefore, some approaches have been initiated such as plant-based omega-3 fatty acid-rich diet. Furthermore, algae oils, plants based oils, different nuts and seeds based oils, legumes, vegetables, grains, and different fruits [47]. Different sources of the omega-3 fatty acid are shown in Table 10.1.

TABLE 10.1   Different Sources of Omega-3 Fatty Acid

Sl. No	Sources	Common Name	Concentration Used	References
1.	Oil source	Olive oil	0.7%	[48, 49]
		Linseed oil	51.9–52.2%	
		Walnut oil	14%	
		Hemp oil	22%	
		Soybean oil	7–10%	
		Mustered oil	6%	
		Pumpkin oil	0.01–0.02	
		Algae oil	-	
		Cod liver oil	-	
		Shark liver oil	-	
2.	Nuts and Seeds	Almond	0.4	[48]
		Flaxseed	22.8	
		Hickory nuts	1.0	
		Peanuts	0.003	
		Walnut black	3.3	
		Walnut, English	6.8	
		Flax	11.4	
3.	Legumes	Beans	0.6	[45, 50]
		Chickpeas	0.1	
		Cowpeas	0.3	
		Lentils	0.1	
		Peas	0.2	
		Soybeans	1.6	
4.	Vegetables	Broccoli	0.1	[48]
		Cauliflower	0.1	
		Kale	0.2	
		Lettuce	0.1	
		Mustard	0.1	
		Purslane	0.4	
		Spinach	0.1	
		Beans	0.3	
		Soybeans	3.2	

TABLE 10.1   *(Continued)*

Sl. No	Sources	Common Name	Concentration Used	References
5.	Grains	Wheat, bran	0.2	[49, 51]
		Wheat, germ	0.7	
		Rice, bran	0.2	
		Oats, germ	1.4	
		Corn, germ	0.3	
		Barley, bran	0.3	
6.	Fruits	Avocados	0.1	[51]
		Raspberries	0.1	
		Strawberry	0.1	
7.	Animals	Tuna	0.2	[50, 52, 53]
		Shark	0.83	
		Egg	0.109	

10.4   MECHANISM OF ACTION OF OMEGA-3 FATTY ACID IN NEURODEGENERATIVE DISORDERS

Alzheimer's is a neurodegenerative disorder noticeable by cognitive and developing diminishing due to impaired learning capability that expressively inhibits with communal and job-related functioning. This kind of behavioral modification abridged stages of brain imitative neurotrophic factor (BDNF), synapsin I, moreover AMP approachable element-binding protein (CREB). It is recognized that BDNF enables synaptic communication and learning capability by moderating synapsin I and CREB. In this situation, one of the foremost counteracting diet sources is omega-3 polyunsaturated fatty acids (i.e., DHA) in the brain, has exposed to be necessary for normal neurological growth, maintenance of learning and remembrance, and neuronal plasticity furthermore control signal transduction and gene expression, and safeguard neurons from death and can decrease cognitive collapse throughout aging and Alzheimer's disease. Consequently, it is believable that select nutritional constituents ingested at the suitable time can be used to regularized stages of BDNF and related synapsin I and CREB, decreased oxidative destruction and counteracted learning disability [54, 55]. The composition of omega-3 fatty acids with the main three chemical constituents, namely, a) alpha-linolenic acid (ALA), b) EPA, c) DHA. These three polyunsaturates have either 3, 5, or 6 double bonds with a carbon structure of 18, 20, or 22 carbon atoms, separately. As with greatest indeed developed fatty acids, totally double bonds are in the *cis*-configuration, in other words, the two hydrogen fragments are with the identical side of the double bond; besides the double bonds are intermittent by methylene bridges, therefore there are two solo bonds among individual pair of head-to-head double bonds [45]. The chemical structures of DHA and EPA are presented in Figure 10.2.

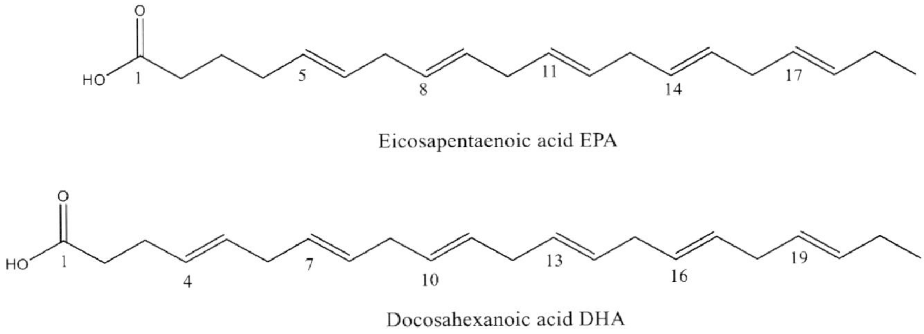

Eicosapentaenoic acid EPA

Docosahexanoic acid DHA

FIGURE 10.2   DHA and EPA chemical structures.

The omega-3 fatty acid has also been exposed to decrease vascular risk, inflammation, and oxidative damage. Existing clinical reports relating the existence of Alzheimer's ailment among mature persons by dissimilar levels of nutritional omega-3 fatty acid consumption, propose that risk of AD is meaningfully reduced amongst those with greater contains fish and omega-3 fatty acid consumption [56].

Unintentional condition and some social traumatic situation develop brain injury that time produces a state of susceptibility that decreases the brain capability to possibility with secondary invectives. The silent information regulator-2 (SIR-2) has been associated with preserving genomic constancy and cellular homeostasis underneath exciting conditions. Aiguo et al. recently reported the novel evidence presenting that mild traumatic brain injury decreases the manifestation of SIR-2 in the hippocampus, in an amount to augmented levels of protein oxidation. Additionally, he shows that dietary supplementation of omega-3 fatty acids that improves protein oxidation was effective to reverse the reduction of the SIR-2 level in wounded rats. Supplementary, he was found that traumatic brain injury reduced ubiquitous mitochondrial creatine kinase (uMtCK – ubiquitous mitochondrial creatine kinase), an enzyme implicated in the energetic regulation of Ca-2 pumps and in the maintenance of Ca-2 homeostasis. Omega-3 fatty acids supplements normalized the levels of uMtCK after lesion [55]. The mechanism of action of Omega-3 fatty acid with neurodegenerative disorders is shown in Figure 10.3.

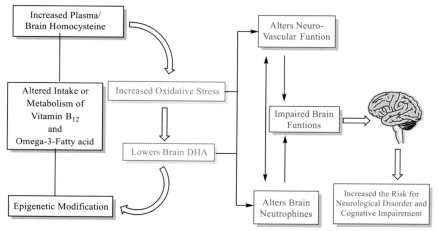

FIGURE 10.3   Mechanism of action of omega-3 fatty acid in different neurodegenerative disorders.

10.5   ACTION OF OMEGA-3 FATTY ACID CONSTITUENTS WITH COMBINATION OR SEPARATELY IN NEURODEGENERATIVE DISORDERS

An omega-3 fatty acid is a chemically active ingredient, which provides the action in a different type of the disease and disorders, with combination or separately. Different applications are discussed below, regarding the omega constituents DHA and EPA. Omega-3 fatty acids having a complementary function in energy uptake and brain utility and that optimum retention of cognitive function, such as ketogenic substrate, a weak precursor to omega-3 PUFA, Stimulation of fatty acid β-oxidation and ketogenesis, neurotransmission, learning, and memory and Brain glucose uptake, etc. [57]. The different Physiological utilities have been recognized to omega-3 FA, including: (1) Transfer of calcium and other substances into and out of cells. (2) Contraction and relaxation of muscles. (3) Operation of clotting and secretion of matters that embrace-gastric enzymes and hormones. (4) Regulators of fertility, cell splitting, and progression [58]. Omega-3 fatty acids are aids to inhibit and treat anxiety and depression with EPA, which is the show's most general anti-depressant activity [59]. DHA is a constituent of the omega-3 fatty acid, which is the most important structural component of eye retinas. It plays a significant role in the inhibition of muscular degeneration, further causes blindness, and vision impairment [60, 61]. Ingesting of sufficient quantity of omega-3 fatty acids with dietary supplements throughout the pregnancy period is helpful for fetus development [62]. People by means of mental disorder often have small plasma levels of omega-3 fats. In this case, it increases the omega-3 fatty acids to improve their symptoms [63]. Omega-3 fatty acids assist to inhibit age-related mental disorder and cognition disorder [64]. Omega-3 intake may reduce the chances of the risk related to the cancer tumors in the brain, and other organs of the human body [65]. Intake of DHA which nothing but a constituent of omega-3 fatty acid, may expand the time span and quality of the sleep [66]. Omega-3 fatty acid serves as an abundant source of the antioxidants (AOs), which provides glow on face, skin, and improves your recognition power [67].

10.6   MARKETED FORMULATION OF OMEGA-3 FATTY ACIDS IN NEURODEGENERATIVE DISORDERS

The omega-3 fatty acid different formulations herein may be provided to a patient in any of a number of pharmaceutically oral dosage forms that are acceptable by the patient. Generally, the omega-3 fatty acids will be orally taken in the form of, tablets, gel, capsules, or pills. However, the administration also is given via any supplementary route where the active ingredients may be capably captivated and used e.g., intravenously, subcutaneously, rectally, vaginally, or topically. This includes herein are pharmaceutical ingredients, comprising pharma-ceutical formulations in a unit dosage form. In these dosage forms, the preparation is divided into appropriately extent unit doses with

TABLE 10.2 Marketed Omega-3 Fatty Acid Formulations

Sl. No.	Product	Description	Dosage	Omega-3 Content	References
1.	Lovaza	Transparent, soft-gelatin capsules packed with light-yellow oil Inactive ingredients include α-tocopherol, gelatin, glycerol, and purified water FDA approval 2004.	Total dose: 4 g/day, taken as: Single 4-g dose (4 capsules) Two 2-g doses (2 capsules BID)	Contains min: 0.9 g of OM3FA ethyl esters: EPA ~0.465 g DHA ~0.375 g	[70]
2.	Vascepa	1-g amber-colored, soft-gelatin capsules. Inactive ingredients include gelatin, glycerin, tocopherol, sorbitol, and purified water. FDA approval in 2012.	Total dose: 4 g/day, taken as: Two 2-g doses (2 capsules BID)	Contains 1 g of IPE and EPA 1 g.	[70]
3.	Epanova	1-g red/brown coated, soft-gelatin capsules. Inactive ingredients include porcine Type A gelatin, glycerol, sorbitol, tocopherol, and purified water. FDA approval in 2014.	Total dose: 2 g/day or 4 g/day, taken as: Single 2-g dose (2 capsules) Single 4-g dose (4 capsules)	Contains min: 1 g of fish-oil-derived free FAs with at least 0.85 g of polyunsaturated FAs, plus multiple OM3FAs from fish oils: EPA 0.550 gDHA 0.2 g	[70]
4.	Omtryg	Transparent capsules filled with light yellow oil Inactive ingredients include α-tocopherol, gelatin, glycerol, and purified water FDA approval 2014.	Total dose: 4 g/day, taken as: Single 4-g dose (4 capsules) Two 2-g doses (2 capsules BID)	Contains min: 0.9 g of OM3FA ethyl esters from oils: EPA ~0.465 g DHA ~0.375 g	[70]
5.	NutrineLife Fish Oil	Omega-3 fish oil soft gelatin capsules in the soft gels with other excipients.	Total dose: 2 Capsules daily	Contains min: 360 mg EPA and 240 mg DHA.	[70]

TABLE 10.2 *(Continued)*

Sl. No.	Product	Description	Dosage	Omega-3 Content	References
6.	Health Kart Omega	Transparent yellowish soft gelatin capsule containing gelatin, lecithin, glycerin Carbohydrate, Fish Lipid Oil	Total dose: Single 4-g dose (4 capsules)	Contains min: Fish oils: 360 mg EPA & 240 mg DHA.	[70]
7.	True Basics Ultra Omega-3 Fatty Acids	Enteric-coated Soft gelatin yellowish capsule True Basics Ultra Omega-3 is with 1250 mg of omega-3 with 460 mg of EPA & 380 mg of DHA.	Total dose: 1.2 gm capsule (1 capsule) daily at bedtime.	Contains min: Omega-3 with 460 mg of EPA & 380 mg of DHA.	[70]
8.	Nutrova Complete Omega-3	Gelatin free capsule with 440 mg of DHA and lemon oil as flavoring agents with HPMC Capsules.	Total Dose: 1–2 capsule per day	Contains min: DHA 440 mg	[70]
9.	Nutricharge Veg Omega	Brownish Soft gelatin capsule containing 420 mg of total omega and 250 mg of Omega-3 nutrients and other excipients.	Total dose: 1 capsule of Nutricharge Veg Omega daily for a long duration after breakfast in the morning.	Contains min: 420 mg of total omega and 250 mg omega-3.	[70]
10.	HealthViva Omega-3	Yellowish gelatin capsules containing EPA and DHA.	Total Dose: 1 capsule of health viva omega-3 daily.	Contains min: 180 mg EPA & 120 mg DHA.	[70]
11.	Medicare Well Flax Oil	Softgel form capsule containing flaxseed oil with omega-3, 6, 9.	As per requirement	Contains: Flaxseed oil	[70]

suitable amounts of the omega-3 fatty acids, from that operative quantity to attain the wanted purpose [68]. Omega-3 fatty acids presently available in the form of a capsule, tablet, liquid, syrup, suspensions, sublingual, candy, and chewable dosage forms of the omega-3 fatty acid preparations. Dosage forms of omega-3 fatty acids available in the EPA and DHA fatty acids are in the triglyceride form, the esterifies form, predominantly the ethyl ester form, in which the fatty acids are in the form of acidic salts [69]. Pharmaceutical dosage forms might comprise excipients like include fillers, surfactants, extenders, binders, surfactants, lubricants, humidifiers, and this excipient should be of adequately high pureness and compatibility by means of lesser toxicity to reduce them appropriate for administration. Excipients should be inactive or it can retain pharmaceutical properties for administration, e.g., oral tablets and capsules, syrups, powders, Suspensions, and other dosage forms which is consist of conservative pharmacy practices. For example, different dosage forms are used like gel, capsules of the omega-3 fatty acid formulation may be in composition with a flavorants, colorants, and preservatives. For special children's use of an oral administration, the amount of omega-3 fatty acid comprised of an oral unit dosage form will be less. Unit dosage forms for children's use provide 10–30 mg/kg omega-3 fatty acid each day in multiple unit dosage forms and for a single dosage form for a pediatric aged group 2 to 6 years, comprises around 50 mg [70–72]. Table 10.2 shows a different marketed formulation containing omega-3 fatty acids.

10.7   FUTURE PROSPECTIVE OF OMEGA-3 FATTY ACID

Currently, due to augmented consciousness of the health paybacks of DHA and EPA, the demand for fish oil increased and fish population day by day decreased. Nevertheless, recommendations to improve in omega-3 fatty acid ingestion will deliberate health threats. During the cooking of the different omega containing foods, it loses fatty acid content during heating and may produce harmful free radicals, which should be considered. An alternative for omega-3 fatty acids such as microalgae which is a rich source of fatty acids, but the production cost of the microalgae is quite high. Plant-based fatty acids similar to marine oil leftovers a big dare, since changing the composition of omega-3 fatty acids in plant seeds to attain a fish oil alternative is the most important objective [73]. Genetic engineering encoding with DHA and EPA biosynthetic pathways with diet grade bacteria like yeast and lactic acid microbes may be used as maintainable and cost-effective substitute sources for the fishes and still effective modern technologies must be applied for that. Nevertheless, transesterification and inter-esterification will leftovers capable techniques of generating omega-3 fatty acids. Meanwhile, omega-3 fatty acids may attention with extra food constituents with the food environment, the impact of like interaction on the nutritive effect of omega-3 fatty acids for further investigation [74–76].

10.8 CONCLUSION

The impact of nutrition has the prospective to significantly effect on the physical role and body uptake. Particular attention has been taken on omega-3 fatty acids, which are obtained from the marine source as well as terrestrial features. This containing DHA and EPA are responsible for the many signaling, various cellular functioning, cellular fluidity, cognition maintenance, and enhancer. Also, controls the different systems related to cognition such as nervous systems, glucose regulation, inflammatory progressions which may be directly or indirectly effects on the cognition system. Animal-based, cells based, fruits based, oils based, seeds based and plant passed, like different types based omega-3 fatty acid sources are reported. The omega-3 fatty acids appear to be amid the utmost important supplements for the enormous variety of the populaces. Additionally, omega-3 fatty acids dietary intake should be increased to overcome the deficiency of the omega in the body to maintain cognition. Likewise, omega-3 fatty acids maintained their own properties after packaging with healthful food other than fish. Nutritional approaches and supplements like omega-3 fatty acids may result in improved regaining, reduced risk of AD, optimal training gain, and increased level of competition enactment. This backing the view about DHA and EPA, which may increase the performance of the in recognition and other related activities. This has the potential benefits regarding neurodegenerative disorders.

KEYWORDS

- **Alzheimer's disease**
- **cognition**
- **docosahexaenoic acid**
- **eicosapentaenoic acid**
- **neurodegenerative**
- **Parkinson's disease**

REFERENCES

1. Hof, P. R., & Mobbs, C. V., (2010). *Handbook of the Neuroscience of Aging.* Academic Press.
2. Przedborski, S., Vila, M., & Jackson-Lewis, V., (2003). Series introduction: Neurodegeneration: What is it and where are we? *The Journal of Clinical Investigation, 111*(1), 3–10.
3. Chen, W. W., Zhang, X., & Huang, W. J., (2016). Role of neuroinflammation in neurodegenerative diseases. *Molecular Medicine Reports, 13*(4), 3391–3396.
4. Czirr, E., & Wyss-Coray, T., (2012). The immunology of neurodegeneration. *The Journal of Clinical Investigation, 122*(4), 1156–1163.
5. Torreilles, F., et al., (1999). Neurodegenerative disorders: The role of peroxynitrite. *Brain Research Reviews, 30*(2), 153–163.
6. Mattson, M. P., (2000). Apoptosis in neurodegenerative disorders. *Nature Reviews Molecular Cell Biology, 1*(2), 120.
7. Martin, J. B., (1999). Molecular basis of the neurodegenerative disorders. *New England Journal of Medicine, 340*(25), 1970–1980.
8. Stefanis, L., Burke, R. E., & Greene, L. A., (1997). Apoptosis in neurodegenerative disorders. *Current Opinion in Neurology, 10*(4), 299–305.
9. Zlokovic, B. V., (2008). The blood-brain barrier in health and chronic neurodegenerative disorders. *Neuron, 57*(2), 178–201.
10. Luchtman, D. W., & Song, C., (2013). Cognitive enhancement by omega-3 fatty acids from child-hood to old age: Findings from animal and clinical studies. *Neuropharmacology, 64,* 550–565.
11. Vöhringer, P. A., et al., (2013). Cognitive impairment in bipolar disorder and schizophrenia: A systematic review. *Frontiers in Psychiatry, 4,* 87.
12. Marazziti, D., et al., (2010). Cognitive impairment in major depression. *European Journal of Pharmacology, 626*(1), 83–86.
13. Reyes, A. N., et al., (2017). Functional impairment and cognitive performance in mood disorders: A community sample of young adults. *Psychiatry Research, 251,* 85–89.

14. Barch, D. M., & Ceaser, A., (2012). Cognition in schizophrenia: Core psychological and neural mechanisms. *Trends in Cognitive Sciences, 16*(1), 27–34.
15. Keefe, R. S., & Harvey, P. D., (2012). Cognitive impairment in schizophrenia, In: *Novel Antischizophrenia Treatments* (pp. 11–37). Springer.
16. Kahn, R. S., & Keefe, R. S., (2013). Schizophrenia is a cognitive illness: Time for a change in focus. *JAMA Psychiatry, 70*(10), 1107–1112.
17. Ayesa-Arriola, R., et al., (2013). Long-term (3-year) neurocognitive effectiveness of antipsychotic medications in first-episode non-affective psychosis: A randomized comparison of haloperidol, olanzapine, and risperidone. *Psychopharmacology, 227*(4), 615–625.
18. Keefe, R. S., et al., (2010). Report from the working group conference on multisite trial design for cognitive remediation in schizophrenia. *Schizophrenia Bulletin, 37*(5), 1057–1065.
19. Buchanan, R. W., et al., (2010). The FDA-NIMH-MATRICS guidelines for clinical trial design of cognitive-enhancing drugs: What do we know 5 years later? *Schizophrenia Bulletin, 37*(6), 1209–1217.
20. Bishop, N. A., Lu, T., & Yankner, B. A., (2010). Neural mechanisms of ageing and cognitive decline. *Nature, 464*(7288), 529.
21. Nelson, P. T., et al., (2011). Alzheimer's disease is not "brain aging": Neuropathological, genetic, and epidemiological human studies. *Acta Neuropathologica, 121*(5), 571–587.
22. Nurk, E., et al., (2007). Cognitive performance among the elderly and dietary fish intake: The Hordaland health study. *The American Journal of Clinical Nutrition, 86*(5), 1470–1478.
23. Gollan, T. H., Salmon, D. P., & Paxton, J. L., (2006). Word association in early Alzheimer's disease. *Brain and Language, 99*(3), 289–303.
24. Maccioni, R. B., et al., (2010). The revitalized tau hypothesis on Alzheimer's disease. *Archives of Medical Research, 41*(3), 226–231.
25. Mamelak, M., (2017). Energy and the Alzheimer brain. *Neurosci Biobehav Rev.*
26. Gillette-Guyonnet, S., et al., (2009). Commentary on "a roadmap for the prevention of dementia II. *Leonthal symposium* 2008." The multi-domain Alzheimer preventive trial (MAPT): A new approach to the prevention of Alzheimer's disease. *Alzheimer's and Dementia, 5*(2), 114–121.
27. Freund-Levi, Y., et al., (2006). Omega-3 fatty acid treatment in 174 patients with mild to moderate Alzheimer disease: Omega study: A randomized double-blind trial. *Archives of Neurology, 63*(10), 1402–1408.
28. Swanson, D., Block, R., & Mousa, S. A., (2012). Omega-3 fatty acids EPA and DHA: Health benefits throughout life. *Advances in Nutrition, 3*(1), 1–7.
29. Su, K. P., et al., (2008). Omega-3 fatty acids for major depressive disorder during pregnancy: Results from a randomized, double-blind, placebo-controlled trial. *Journal of Clinical Psychiatry, 69*(4), 644.
30. Lazzarin, N., et al., (2009). Low-dose aspirin and omega-3 fatty acids improve uterine artery blood flow velocity in women with recurrent miscarriage due to impaired uterine perfusion. *Fertility and Sterility, 92*(1), 296–300.

31. Smith, G. I., et al., (2010). Dietary omega-3 fatty acid supplementation increases the rate of muscle protein synthesis in older adults: A randomized controlled trial. *The American Journal of Clinical Nutrition, 93*(2), 402–412.

32. Conquer, J. A., et al., (2000). Fatty acid analysis of blood plasma of patients with Alzheimer's disease, other types of dementia, and cognitive impairment. *Lipids, 35*(12), 1305–1312.

33. Susanne Krauss-Etschmann, Rania Shadid, Cristina Campoy, et al., (2007). Effects of fish-oil and folate supplementation of pregnant women on maternal and fetal plasma concentrations of docosahexaenoic acid and eicosapentaenoic acid: A European randomized multicenter trial. *The American Journal of Clinical Nutrition, 85*(5), 1392–1400.

34. Serhan, C. N., Chiang, N., & Van, D. T. E., (2008). Resolving inflammation: Dual anti-inflammatory and pro-resolution lipid mediators. *Nature Reviews Immunology, 8*(5), 349.

35. Neff, L. M., et al., (2010). Algal docosahexaenoic acid affects plasma lipoprotein particle size distribution in overweight and obese adults. *The Journal of Nutrition, 141*(2), 207–213.

36. Freund-Levi, Y., et al., (2008). Omega-3 supplementation in mild to moderate Alzheimer's disease: Effects on neuropsychiatric symptoms. *International Journal of Geriatric Psychiatry, 23*(2), 161–169.

37. Prince, M., et al., (2016). *World Alzheimer Report 2016: Improving Healthcare for People Living with Dementia: Coverage, Quality, and Costs Now and in the Future.*

38. Jicha, G. A., & Markesbery, W. R., (2010). Omega-3 fatty acids: Potential role in the management of early Alzheimer's disease. *Clinical Interventions in Aging, 5*, 45.

39. Devore, E. E., et al., (2009). Dietary intake of fish and omega-3 fatty acids in relation to long-term dementia risk. *The American Journal of Clinical Nutrition, 90*(1), 170–176.

40. Roberts, R. O., et al., (2010). Polyunsaturated fatty acids and reduced odds of MCI: The mayo clinic study of aging. *Journal of Alzheimer's Disease, 21*(3), 853–865.

41. Cherubini, A., et al., (2007). Low plasma N-3 fatty acids and dementia in older persons: The in CHIANTI study. *The Journals of Gerontology Series A: Biological Sciences and Medical Sciences, 62*(10), 1120–1126.

42. Heude, B., Ducimetière, P., & Berr, C., (2003). Cognitive decline and fatty acid composition of erythrocyte membranes: The EVA study. *The American Journal of Clinical Nutrition, 77*(4), 803–808.

43. Chiu, C. C., et al., (2012). Associations between n-3 PUFA concentrations and cognitive function after recovery from late-life depression. *The American Journal of Clinical Nutrition, 95*(2), 420–427.

44. Johnson, M., & Bradford, C., (2014). Omega-3, omega-6, and omega-9 fatty acids: Implications for cardiovascular and other diseases. *J. Glycomics Lipidomics, 4*(123), 2153–0637.1000123.

45. Kaur, M., Basu, S., & Shivhare, U. S., (2015). Omega-3 fatty acids: Nutritional aspects, sources, and encapsulation strategies for food fortification. *Direct Research Journal of Health and Pharmacology (DRJHP), 3*(1), 12–31.

46. Garg, M., et al., (2006). Means of delivering recommended levels of long chain n-3 polyunsaturated fatty acids in human diets. *Journal of Food Science, 71*(5), R66–R71.

47. Lenihan-Geels, G., Bishop, K., & Ferguson, L., (2013). Alternative sources of omega-3 fats: Can we find a sustainable substitute for fish? *Nutrients, 5*(4), 1301–1315.

48. Kris-Etherton, P. M., Harris, W. S., & Appel, L. J., (2003). Fish consumption, fish oil, omega-3 fatty acids, and cardiovascular disease. *Arteriosclerosis, Thrombosis, and Vascular Biology, 23*(2), e20–e30.

49. Tur, J., et al., (2012). Dietary sources of omega-3 fatty acids: Public health risks and benefits. *British Journal of Nutrition, 107*(S2), S23–S52.

50. Kolanowski, W., & Laufenberg, G., (2006). Enrichment of food products with polyunsaturated fatty acids by fish oil addition. *European Food Research and Technology, 222*(3–4), 472–477.

51. Kris-Etherton, P. M., Harris, W. S., & Appel, L. J., (2002). Fish consumption, fish oil, omega-3 fatty acids, and cardiovascular disease. *Circulation, 106*(21), 2747–2757.

52. Maurice, D., (1994). Dietary fish oils: Feeding to produce designer eggs. *Feed Manag., 45*, 29–32.

53. Mataix, J., (2003). *Tabla De Composición De Alimentos (Food composition tables).* Granada: University of Granada.

54. Wu, A., Ying, Z., & Gomez-Pinilla, F., (2004). Dietary omega-3 fatty acids normalize BDNF levels, reduce oxidative damage, and counteract learning disability after traumatic brain injury in rats. *Journal of Neurotrauma, 21*(10), 1457–1467.

55. Wu, A., Ying, Z., & Gomez-Pinilla, F., (2007). Omega-3 fatty acids supplementation restores mechanisms that maintain brain homeostasis in traumatic brain injury. *Journal of Neurotrauma, 24*(10), 1587–1595.

56. Lim, W. S., et al., (2006). Omega-3 fatty acid for the prevention of dementia. *Cochrane Database of Systematic Reviews*, p. 1.

57. Freemantle, E., et al., (2006). Omega-3 fatty acids, energy substrates, and brain function during aging. *Prostaglandins, Leukotrienes and Essential Fatty Acids, 75*(3), 213–220.

58. MacLean, C. H., et al., (2005). *Effects of Omega-3 Fatty Acids on Cognitive Function with Aging, Dementia, and Neurological Diseases: Summary, in AHRQ Evidence Report Summaries.* Agency for Healthcare Research and Quality (US).

59. Jazayeri, S., et al., (2008). Comparison of therapeutic effects of omega-3 fatty acid eicosapentaenoic acid and fluoxetine, separately and in combination, in major depressive disorder. *Australian and New Zealand Journal of Psychiatry, 42*(3), 192–198.

60. Lim, L. S., et al., (2012). Age-related macular degeneration. *The Lancet, 379*(9827), 1728–1738.

61. Merle, B. M., et al., (2014). Circulating omega-3 fatty acids and neovascular age-related macular degeneration. *Investigative Ophthalmology and Visual Science, 55*(3), 2010–2019.

62. Helland, I. B., et al., (2003). Maternal supplementation with very-long-chain n-3 fatty acids during pregnancy and lactation augments children's IQ at 4 years of age. *Pediatrics, 111*(1), e39–e44.

63. Benton, D., (2007). The impact of diet on anti-social, violent, and criminal behavior. *Neuroscience and Biobehavioral Reviews, 31*(5), 752–774.

64. Canhada, S., et al., (2018). Omega-3 fatty acids' supplementation in Alzheimer's disease: A systematic review. *Nutritional Neuroscience, 21*(8), 529–538.

65. Fabian, C. J., Kimler, B. F., & Hursting, S. D., (2015). Omega-3 fatty acids for breast cancer prevention and survivorship. *Breast Cancer Research, 17*(1), 62.

66. Hansen, A. L., et al., (2014). Fish consumption, sleep, daily functioning, and heart rate variability. *Journal of Clinical Sleep Medicine, 10*(05), 567–575.

67. Spencer, E. H., Ferdowsian, H. R., & Barnard, N. D., (2009). Diet and acne: A review of the evidence. *International Journal of Dermatology, 48*(4), 339–347.

68. Salem, Jr. N., & Eggersdorfer, M., (2015). Is the world supply of omega-3 fatty acids adequate for optimal human nutrition? *Current Opinion in Clinical Nutrition and Metabolic Care, 18*(2), 147–154.

69. Brinton, E. A., & Mason, R. P., (2017). Prescription omega-3 fatty acid products containing highly purified eicosapentaenoic acid (EPA*). Lipids in Health and Disease, 16*(1), 23.

70. Bradberry, J. C., & Hilleman, D. E., (2013). Overview of omega-3 fatty acid therapies. *Pharmacy and Therapeutics, 38*(11), 681.

71. Andrews, K. W., et al., (2017). *USDA Dietary Supplement Ingredient Database Release 4.0 (DSID-4).*

72. Adarme-Vega, T. C., Thomas-Hall, S. R., & Schenk, P. M., (2014). Towards sustainable sources for omega-3 fatty acids production. *Current Opinion in Biotechnology, 26,* 14–18.

73. Marangoni, F., & Poli, A., (2013). n-3 fatty acids: Functional differences between food intake, oral supplementation and drug treatments. *International Journal of Cardiology, 170*(2), S12–S15.

74. Ofosu, F. K., et al. (2017). Current trends and future perspectives on omega-3 fatty acids. *Res. Rev. J. Biol., 5,* 11–20.

75. Lazic, M., et al., (2014). Reduced dietary omega-6 to omega-3 fatty acid ratio and 12/15-lipoxygenase deficiency are protective against chronic high fat diet-induced steatohepatitis. *PloS One, 9*(9), e107658.

76. Banaschewski, T., et al., (2018). Supplementation with polyunsaturated fatty acids (PUFAs) in the management of attention deficit hyperactivity disorder (ADHD). *Nutrition and Health, 24*(4), 279–284.

CHAPTER 11

Thymoquinone: Therapeutic Potential and Molecular Targets

ABUZER ALI,[1] MUSARRAT HUSAIN WARSI,[1] WASIM AHMAD,[2] MOHD. AMIR,[3] NIYAZ AHMAD,[3] AMENA ALI,[1] and ABUTAHIR[4]

[1]College of Pharmacy, Taif University, Haweiah, Taif, Saudi Arabia, Tel.: +966599131866, E-mail: abuzerali007@gmail.com (A. Ali)

[2]Department of Pharmacy, Mohammad Al-Mana College for Medical Sciences, Dammam, Saudi Arabia

[3]College of Clinical Pharmacy, Imam Abdulrahman Bin Faisal University, Dammam, Saudi Arabia

[4]Raghukul College of Pharmacy, Sarvar, Bhopal, Madhya Pradesh, India

ABSTRACT

Naturally, thymoquinone (TQ) is present in the highest amount in *N. sativa* seed which is popular among several traditional systems of medicine. Chemically, TQ is 2-isopropyl-5-methyl-1,4-benzoquinone, belongs to the category of monoterpene. It is extremely lipophilic in nature with poor solubility and bioavailability. TQ possesses multiple therapeutic properties like immunomodulatory, anti-oxidant, anti-inflammatory, anti-histaminic, anti-hypertensive, anti-microbial, anti-Alzheimer, anti-Parkinson, anti-arthritis, anti-diabetic, and anti-tumor effects. It also displays synergistic effects with many conventional drugs. TQ modulates various receptors, transcription factors, enzymatic, and cell signaling pathways to exhibit its pharmacological effects. Pharmacological properties, high therapeutic index, pharmacokinetics, lipophilicity, efficacy, and low toxicity profile represent TQ as a capable molecule for drug development. This chapter summarizes significant research findings related to TQ with its pharmacological potential and major molecular targets including its importance in human health.

11.1 INTRODUCTION

Currently, there has been a shift in global trend from synthetic to plant-based medicine, called 'Return to Nature.' Plants with medicinal values have been acknowledged for centuries and are well known around the globe as a rich source of medicinal agents for the prevention and cure of diseases. The basis for recognition and preference is faith that herbal prod-ucts are harmless and safe [1]. The recognition of natural products (NPs) as non-toxic, with lesser side effects and easily accessible at reasonable costs, made them demanding worldwide for plant-based health products, food supplements, pharmaceuticals, and cosmetics. At present, there is a reinforcement of interest with plant-based medicine due to the increasing awareness of the health hazards related to indiscriminate use of modern medicines [2]. The indigenous knowledge of botanicals has been conveyed from generation to generation throughout the world and has extensively contributed to the development of diverse traditional systems of medicine. The utilization of botanicals as medicines has implicated the isolation and characterization of pharmacologically active compounds. It is anticipated that approximately 2,50,000 plant species are found worldwide. They persistently offer natural compounds with the power of healing and for the discovery and development of novel drugs [3].

The medicinal property of plants is due to the synthesized chemical compounds by them in their different parts. Most of the natural compounds like: terpenes, glycosides, alkaloids, steroids, saponins, flavonoids, polyphenols, coumarins, and tannins possess significant pharmacological potentials [4]. These natural compounds keep different groups/moieties as pharmacophores having the potential for biological interaction. Among various moieties, one of the classes is quinones including anthraquinones, naphthoquinone, and benzoquinone. TQ is a benzoquinone having multitargeted therapeutic potential that helps to retain constant attention from researchers around the globe [5]. Thus, in this chapter, the authors reviewed and mentioned significant research findings related to TQ with its pharmacological potential and major molecular targets including its importance in human health.

11.2   OCCURRENCE OF TQ IN PLANTS

Along with *Nigella sativa*, TQ is found in trace amount in *N. arvensis* L. seeds belonging to the same family of Ranunculaceae. TQ has been detected in different genera like Agastache, Coridothymus, Monarda, Mosla, Origanum, Satureja, Thymbra, and Thymus of Lamiaceae family. The occurrence of TQ has also been confirmed in the Cupressaceae family of genera Tetraclinis, Cupressus, and Juniperus. TQ exists in a glycosidic form in Cupressus and Juniperus gerena. However, TQ occurs together with dithymoquinone (DTQ) (dimeric form) and thymohydroquinone (THQ) (reduced form) in various species of plants. Like TQ, THQ is found in various genera such as Coridothymus, Origanum, Monarda, Mosla, Satureja, and Thymus of Lamiaceae family. Its occurrence has also established in the genera Nigella and Tetraclinis. While DTQ has noticed in a few species of Ranunculaceae and Lamiaceae families [6, 7].

11.3   IMPORTANCE OF *NIGELLA SATIVA*

Naturally, TQ is present in the highest amount in *N. sativa* seed which is popular among several traditional systems of medicine. *N. sativa* or black cumin is a promising medicinal plant that comes under the Ranunculaceae family and categorized as "Generally Recognized as Safe" by United States Food and Drug Administration (FDA). The seed and oil of this plant have a famous and spiritual history for various health conditions. The benefits of *N. sativa* seed are mentioned in the Bible and in Islamic scriptures as "seed of blessing" [4–6]. Geographically, *N. sativa* is distributed to Southern Europe, North Africa, and Southwest Asia and cultivated in many countries including Syria, Saudi Arabia, Turkey, India, and Pakistan. *N. sativa* seeds have been used as both food and medicine and mentioned by Avicenna in "The Canon of Medicine" [8]. Ethnobotanically seeds are used in the treatment of cough, bronchitis, asthma, fever, headache, dizziness, inflammation, eczema, influenza, hypertension, rheumatism, diabetes, nervous disorders, liver, gastrointestinal, and kidney problems, cancer, and related inflammatory diseases [9, 10]. *N. sativa* seeds are reported to contain diverse chemical components such as amino acids, proteins, carbohydrates, organic acids, crude fibers, minerals, vitamins, alkaloids, saponins, terpenes either in the fixed or essential oil (EO) [6,

11]. Many compounds have been isolated and reported from *N. sativa* seeds, out of which TQ (30%–48%) is the most abundant and important bioactive compound. Along with TQ, seeds also contain DTQ, THQ, α-pinene, carvacrol, carvone, p-cymene, 4-terpineol, sesquiterpene longifolene, t-anethol, thymol, citronellol, and limonene. Fixed oil from seeds mainly consists of linoleic, oleic, dihomolinoleic, eicosadienoic, stearic, and palmitic acid. *N. sativa* seeds also reported containing stigmasterol, α-sitosterol, nigellicimine, nigellicimine-N-oxide, nigellicine, nigellidine, and alpha-hederin [4, 8].

11.4   CHEMICAL AND PHYSICAL PROPERTIES OF TQ

Chemically, TQ is 2-methyl-5-propan-2-ylcyclohexa-2,5-diene-1,4-dione or 2-isopropyl-5-methyl-1,4-benzoquinone, belongs to the category of monoterpene. It is a white to yellow crystalline powder with melting point 44–45°C. Its molecular weight is 164.2 g/mol and represented by the molecular formula, $C_{10}H_{12}O_2$. It is hydrophobic in nature thus, soluble in alcohol and organic solvents, sparingly soluble in aqueous buffers, and insoluble in water. TQ is highly thermolabile in nature and its partition coefficient is 2.54 (LogP) [6]. Due to these qualities, the pharmaceutical development of TQ into different dosage forms is difficult. Salem [9], proposed that TQ shows keto-enol tautomerism, where the keto-form is considered to be the main configuration and responsible for pharmacological properties [9]. However, the chemical structure of TQ does not allow the possibility of keto-enol tautomerism, but in reality, TQ itself misinterpreted as keto form and its reduced form THQ ($C_{10}H_{14}O_2$) is misunderstood as enol form by the researchers [12]. The absorbance maxima (λ_{max}) for TQ range 254–257 nm. It is highly sensitive to light and can be degraded on exposure even for a short duration. Further, it is unstable at alkaline pH and its stability decreases with increasing pH [6].

11.5   PHARMACOKINETICS OF TQ

The extremely lipophilic nature of TQ is the reason for its poor solubility and bioavailability. Oral administration of TQ might cause its biotransformation by quinone reductase (liver enzyme) that reduces it into hydroquinone. However, one study described that different dose range of TQ

(4–50 mg/kg i.p.) did not make any changes in the biochemical parameters in mice. Intraperitoneal administration of TQ in mice reported LD_{50} of 90.3 mg/kg while LD_{50} of 10 mg/kg was reported in rats [13–15]. Various studies displayed that on oral administration, TQ (10–100 mg/kg) did not exhibit toxic or lethal effects on experimental animals [16–20]. A study showed its clearance after intravenous administration as 7.19 ml/kg/min, the volume of distribution at steady state as 700.90 ml/kg, and absorption half-life ($T_{1/2}$) about 217 min. TQ was quickly removed from the plasma. However, TQ-protein binding in the rabbit was about 99.19% while in human plasma was 98.99%. Results displayed that the TQ absorb slowly after oral administration and eliminate quickly from the body [6, 21].

11.6 PHARMACOLOGICAL POTENTIAL OF TQ

TQ possesses multiple therapeutic properties such as anti-oxidant, anti-inflammatory, immunomodulatory, anti-histaminic, anti-hypertensive, anti-microbial, anti-tumor effects. The earlier research findings proved that TQ has potential as anti-Alzheimer, anti-Parkinson, anti-arthritis, anti-diabetic, and possesses neuroprotective, hepatoprotective, gastroprotective, and nephroprotective chemotherapeutic activities. It has shown positive effects against apoptosis and oxidative stress including respiratory, cardiovascular, and urinary diseases. TQ also exhibited defensive action against various types of cancers and displayed antitumor effects against several cancer cells (CRC) including pancreas, osteosarcoma, lung, ovarian, colon, and myeloblastic leukemia [6, 22, 23].

11.7 ANTI-OXIDATIVE, ANTI-INFLAMMATORY, AND IMMUNOMODULATORY POTENTIAL OF TQ

Anti-oxidant enzymes in the human body such as superoxide dismutase (SOD), glutathione peroxidize (GTPx), glutathione transferase (GST), glutathione reductase (GR), and catalase (CAT) form defense mechanisms to protect against cell damage induced by reactive oxygen species (ROS). TQ possesses potent scavenging activity against ROS like hydrogen peroxide (H_2O_2), superoxide anion ($O2^{-\cdot}$), hydroxyl radical ($\cdot OH$), and peroxyl radical ($ROO^{\cdot}$) [16, 24]. Redox properties of TQ are mainly due to its quinone moiety. It can easily cross-physiological barriers and enter subcellular compartments that support its radical scavenging effects. It is a powerful inhibitor of iron-dependent microsomal lipid peroxidation (LPO). TQ has the potential to induce the expression and/or activity of SOD, GST, GR, and GTPx [25–29]. It displayed protective action on the kidney against mercuric chloride, gentamicin, vancomycin, doxorubicin, and cisplatin-induced nephrotoxicity. TQ exhibited protecting action on the liver from aflatoxin B1 (AFB1), carbon tetrachloride, acetaminophen, and cyclophosphamide-induced hepatic toxicity. TQ (5 mg/kg) pre-treatment helped to reinstate the condition in 1,2-dimethylhydrazine-induced oxidative stress in colon tumors. It showed anti-oxidative and anti-inflammatory actions in the hippocampal neurodegeneration rat model on prolonging toluene exposure [6].

Inflammation is a complex body immune response to harmful stimuli facilitated by cyclooxygenase (COX) and lipoxygenase (LOX) enzymes. COX is responsible for the production of prostaglandins (PGs) while LOX generates leukotrienes (LTs) [24]. TQ exhibited anti-inflammatory action through inhibition of both COX and LOX enzymes and the associated generation of eicosanoids (thromboxane, PG, LT) [30]. It exhibited potent anti-inflammatory action in an asthmatic murine model and showed inhibition of aggregation of inflammatory cells in bronchoalveolar lavage (BAL) fluid and lung tissues. It also showed an anti-inflammatory effect in an allergic asthma mouse model, where it significantly exhibited the inhibition of allergen-induced lung eosinophilic inflammation and displayed inhibitory effects on IL-4, IL-5, and IL-13 with slight inducing effect on the production of interferon-gamma (IFN-γ) in the BAL fluid [31, 32]. In the in-vivo model, TQ displayed down-regulation in 5-LOXexpressionand reduced the production of LTB4 and LTC4 and showed inhibition of COX-2 expression and PGD2 [33, 34]. Hayat and colleagues [35] demonstrated the potential of TQ against allergic conjunctivitis. On administration, it showed a reduction in the symptoms of allergic conjunctivitis by decreasing the levels of IgE, cytokines, histamine, and recruitment of eosinophils in mice [35].

TQ was effective against in-vivo arthritic animal models [36–38] and autoimmune encephalomyelitis [39, 40]. TQ showed an anti-arthritic effect in collagen-induced arthritis by suppressing the level of nitric oxide (NO) and myeloperoxidase while improving the activity of CAT, GSH, and SOD. Furthermore, it significantly decreased pro-inflammatory mediators

(IL-1β, IL-6, IFN-γ, PGE2, and tumor necrosis factor (TNF-α) levels and enhanced the level of IL-10. TQ showed similar action like methotrexate in adjuvant-induced arthritis [37] and significantly determined to reduce the levels of IL-1β, TNF-α, metalloproteinase-13, COX-2 and PG with a decrease in arthritis scoring and bone histology [38]. Experimental autoimmune encephalitis or experimental allergic encephalitis (EAE) is a commonly used animal model for the human inflammatory demyelinating disease (multiple sclerosis). TQ improved the GSH level without showing perivascular inflammation and any disease symptoms in EAE rats when compared with untreated groups. It also inhibited NF-κB activation in encephalomyelitis in animals and reported to exhibit therapeutic effects in EAE animals [39, 40]. It showed inhibition of IL-6 signaling which proved its action against EAE, allergic lung inflammation, allergic asthma, experimental colitis, and rheumatoid arthritis. TQ showed immunomodulatory effect by suppressing cytokines IL-1β and IL-8 levels in mixed lymphocyte cultures. It also improved the activity of antigen-specific CD8T cells and provides protection against tumor growth [6].

11.8 ANTICANCER AND CYTOTOXIC POTENTIAL OF TQ

Many in vivo and in vitro studies have been reported the efficacy of TQ against various types of cancer. TQ showed significant anticancer effect against brain [41, 42], breast [43, 44], lung [45, 46], gastric [47, 48], bladder [49, 50], colon [51, 52], prostate [53] and bone cancer [54]. TQ has demonstrated its anticancer activity through different mechanisms of action. It affects various biological pathways that involve cell proliferation and cell cycle regulation, apoptosis, angiogenesis, and metastasis [55]. It showed inhibition of carcinogenesis in various experimental animal models. It arrests the growth of numerous cancer cell lines in vitro and xenograft tumors in vivo. TQ also exhibited synergistic effects in combination with other anticancer drugs [56]. It showed a preventive role against the growth of medulloblastoma (MB) cells by triggering G2M cell cycle arrest and induction of apoptosis [42]. It also caused telomere shortening, DNA damage, and apoptosis in human glioblastoma cells [57]. TQ with piperine exhibited synergistic effects against breast cancer in vitro and in vivo. A combination of these two constituents showed antineoplastic activity against breast cancer in mice with a significant decrease in tumor size by suppressing vascular endothelial growth factor (VEGF) expression and increasing serum IFN-γ levels. Further, this combination showed angiogenesis inhibition, increasing apoptosis, and shifting immune response toward T-helper1 cell response [58]. In another study, a combination of TQ with resveratrol displayed a significant decrease in tumor size in Balb/C mice. Combination therapy exerted induction in geographic necrosis and apoptosis while the reduction in VEGF expression [43]. Ganji-Harsini et al. [59] reported the anticancer potential of TQ in combination with tamoxifen on estrogen-positive MCF-7 and estrogen negative MDA-MB-231 human breast cancer cell lines [59]. Further, it showed cytotoxicity and apoptotic effects in triple-negative breast cancer cell lines and inhibited tumor growth in vivo. TQ and paclitaxel combination restricted cancer growth in vitro and in vivo both. TQ alone and TQ-paclitaxel combination upregulated tumor suppressor genes like p21, Brca1, and Hic1 [60]. TQ-loaded nanogels exhibited cytotoxic activity and significantly prevented the proliferation of human breast adenocarcinoma cell line (MCF7) [61, 62].

Roepke and co-workers [54] assessed the antiproliferative and pro-apoptotic effects of TQ in two human osteosarcoma cell lines. TQ showed a reduction in cell survival more significantly, in p53-null MG63 cells than in p53-mutant MNNG/HOS cells [54]. In another study, antitumor, and anti-angiogenic, effects of TQ were investigated on osteosarcoma in vitro and in vivo. It displayed higher growth inhibition and apoptosis in the human osteosarcoma cell line (SaOS-2) than control. TQ showed significant downregulation of NF-κB DNA-binding activity, X-linked inhibitor of apoptosis, survivin, and VEGF while upregulation of cleaved caspase-3 and second mitochondrial-derived activator of caspases (SMAC) expression in osteosarcoma cells [63]. Ng et al. [64] reported the cytotoxic effects of TQ-loaded nano-carrier on cervical cancer cell lines (HeLa and SiHa). It caused the induction of apoptosis and cell cycle arrest in breast cancer cell line (MDA-MB-231) [64]. In another study, TQ showed anticancer potential against HeLa and SiHa human cervical CRC by inhibiting proliferation [65, 66]. Its treatment exhibited promising effects on lung CRC (A549) via inhibiting proliferation, migration, and invasion, and caused the reduction of phosphorylation of extracellular signal-regulated protein kinases 1 and 2 (ERK1/2). It also presented the upregulation of Bax and downregulation of Bcl2 proteins with an increase in the Bax/Bcl2 ratio [45, 67]. Multiple studies conducted by different investigators reported the effectiveness of TQ against colon and colorectal cancer both *in vitro* and *in vivo* [52, 68–70]. TQ treatment significantly reduced the number of tumors and large aberrant crypts foci. It caused a significant reduction in the level of Wnt proteins, β-catenin, heat shock protein 90 (HSP-90), COX-2, NF-κB, VEGF, inducible nitric oxide synthase (iNOS) while showed significant induction in anti-tumorigenesis proteins like cyclin-dependent kinase inhibitor 1A (CDKN-1A), dickkopf-related protein 1 (DKK-1), transforming growth factor, beta receptor II (TGF-β/RII), TGF-β1 and mothers against decapentaplegic homolog 4 (SMAD4) [62, 68].

11.9   CARDIOPROTECTIVE POTENTIAL OF TQ

Many research studies have been investigated the effectiveness of TQ against cardiovascular diseases (CVDs) including atherosclerosis [5, 22]. TQ exhibited cardioprotective effects in isoproterenol-induced myocardial injury in vivo. Treatment of TQ improved antioxidants (AOs) level and protected cardiomyocytes with inhibition of LPO and reduction in proinflammatory cytokines level [71]. In two different studies, TQ showed cardioprotective action against cyclophosphamide and doxorubicin-induced cardiotoxicity, where it improved the biological alterations caused by both cardiotoxic agents [72, 73]. Nader et al. [74] demonstrated the potential of TQ against the risk of atherosclerosis in vivo. Its administration with a cholesterol-enriched diet caused a significant reduction in the level of total cholesterol (TC), low-density lipoprotein-cholesterol (LDL-C), triglycerides (TG), and thiobarbituric acid-reactive substances while improved the level of high-density lipoprotein-cholesterol (HDL-C) and glutathione content [74]. In another study, TQ treatment to hypercholesterolemic rabbits displayed inhibition in the progress of atherosclerosis, reduction in oxidative stress, and lipid profiles. It caused a reduction in protein carbonyl and the levels of TC/HDL-C, LDL-C, TG, and malondialdehyde (MDA) [74]. Ragheb and co-workers [76] investigated TQ against atherosclerosis induced by high cholesterol diet and cyclosporine. It showed a reduction in oxidative stress, aortic MDA, and prevented atherogenesis [76]. Antioxidant and hypolipidemic activities of TQ were examined in atherosclerotic rats. TQ exhibited improvement in cardiovascular risks through an increase in the arylesterase enzyme while a decrease in β-hydroxy β-methylglutaryl-CoA (HMG-CoA) reductase activity. It also blocked the formation of MDA and displayed the modification in the level of LDL [22, 77].

11.10   ANTIDIABETIC POTENTIAL OF TQ

Diabetes mellitus (DM) associated with CVDs is dangerous for human health [78]. ROS takes part as an essential element in the development of DM. Several studies reported that *N. sativa* and its components are effective and potent against diabetes [22]. TQ displayed anti-diabetic action by protecting the β-cells from injury and caused a reduction in ROS level [79–81]. Al-Wafai et al. [82] investigated the anti-diabetic potential of TQ in the STZ induced diabetic rats. It exhibited an alteration in the function of the COX-2 enzyme, a reduction in the MDA, and improvement in the SOD levels in the pancreatic tissue of diabetic rats [82]. TQ treatment inhibited the synthesis of gluconeogenic enzymes in the gestational diabetic hamster and showed a decrease in blood glucose levels. On the other hand, the administration of TQ protected the offspring of maternal diabetic mice from obesity and diabetes. It restored the levels of insulin, blood glucose, free radicals, IL-1b, IL-6, TNF-a, and lymphocyte proliferation in offspring [83, 84]. In another study, TQ displayed anti-diabetic potential in hyperglycemic rats against STZ-nicotinamide induced diabetes. It reduced the activities of glucose-6-phosphatase and fructose-1, 6-bisphosphatase, and showed improvement in insulin and reduction in blood glucose level [85]. Administration of TQ in diabetic rats displayed improvement in pancreatic insulin release with a decrease in blood glucose levels. It ameliorated the toxic effects of STZ, like the segregation of nucleoli, DNA damage, fragmentation, and vacuolization of mitochondria by modifying oxidative stress [22, 86].

11.11   ANTIMICROBIAL POTENTIAL OF TQ

Kouidhi et al. [87] investigated the antibacterial activity of TQ against oral pathogens. TQ exhibited a synergic effect with both tetracycline and benzalkonium chloride (4-fold potentiation). It also blocked the 4,6-diamidino-2-phenylindole (DAPI) efflux activity in a concentration-dependent manner [87]. *N. sativa* EO and TQ were evaluated against several clinical cariogenic bacteria by Harzallah and co-workers [88].

The EO containing TQ displayed significant antibacterial activity against *Streptococcus mitis*, *S. constellatus*, *S. mutans*, and *Gemella haemolysans* while TQ alone was effective against all the strains [88]. In another report, it showed preventive effect against the formation of bacterial biofilm especially towards Gram-positive cocci [89]. Halawani [90] investigated the antibacterial potential of TQ and THQ against *S. aureus*, *Escherichia coli*, *Shigella flexneri*, *Pseudomonas aeruginosa*, *Salmonella typhimurium*, and *Salmonella enteritidis*. TQ was more active than THQ against all the strains and showed synergism with other antibiotics against *S. aureus* [90]. Aljabre et al. [91] investigated the antifungal activity of *N. sativa* extract and TQ on various species of dermatophytes. TQ and *N. sativa* extract exhibited a greater minimum inhibitory concentration (MIC) than griseofulvin [91]. Further, TQ treatment displayed inhibitory action against parasites such as *Entamoeba histolytica* causing amoebiasis and *Giardia lamblia* causing giardiasis [92]. Umar and his colleagues [93] reported the antiviral and immunomodulatory effects of TQ in combination with curcumin. Combining the effect of both the drugs showed immunomodulatory action against H9N2 avian influenza virus in vivo by improving the antibody level in the blood, while enhanced cytokine gene expression proposed antiviral behavior of the combination [5, 93].

11.12 CONCLUSION

This review highlights pharmacological effects and molecular targets of TQ and enumerates its benefits against various diseases and complications such as oxidative stress, inflammation, cancer, diabetes, cardiovascular problems, and microbial infections. TQ is a dietary component and found in the highest amount in *N. sativa* seeds which are popular among several traditional systems of medicine. Multiple therapeutic potential and low toxicity make *N. sativa* and TQ popular worldwide to be used as a dietary supplement and attracted scientific attention for pharmacological and drug development approaches. Pharmacological properties, high therapeutic index, pharmacokinetics, lipophilicity, efficacy, and low toxicity profile represent TQ as a capable molecule for drug development. It also displayed synergistic effects with many conventional drugs and provides a justification for its use in multifactorial pathogenesis. Further research studies should continue in order to understand the precise molecular mechanism of TQ and to develop its potent analogs. Regardless of its therapeutic efficacy in animal models, long-term clinical trials should be conducted to bring TQ into the clinic.

KEYWORDS

- anticancer
- bronchoalveolar lavage
- dithymoquinone
- *Nigella sativa*
- thymohydroquinone pharmacological effect
- thymoquinone

REFERENCES

1. Sharma, A., Shanker, C., Tyagi, L. K., Singh, M., & Rao, C. V., (2008). Herbal medicine for market potential in India: An Overview. *Academic Journal of Plant Sciences, 1*(2), 26–36.
2. Satakopan, S., (1994). Pharmacopeial standards for ayurvedic, siddha and unani drugs. In: *Proceedings of WHO Seminar on Medicinal Plants and Quality Control of Drugs Used in ISM*. Ghaziabad.
3. Jachak, S. M., & Saklani, A., (2007). Challenges and opportunities in drug discovery from plants. *Current Science, 92*(9), 1251–1257.
4. Anjali, S., Iqbal, A., Sohail, A., Mohammad, Z. A., Zeenat, I., & Farhan, J. A., (2012). Thymoquinone: Major molecular targets, prominent pharmacological actions, and drug delivery concerns. *Current Bioactive Compounds, 8*(4), 334–344.
5. Sameer, N. G., Chaitali, P. P., Prashant, R. G., Chandragouda, R. P., Umesh, B. M., Charu, S., Sandhya, P. T., & Shreesh, K. O., (2017). Therapeutic potential and pharmaceutical development of thymoquinone: A multi-targeted molecule of natural origin. *Frontiers in Pharmacology, 8*, 656, 1–19.
6. Sara, D., Ali, B. P., Abasalt, H. C., & Sajjad, S., (2015). Thymoquinone and its therapeutic potentials. *Pharmacological Research, 95–96*, 138–158.
7. Jan, T., Miroslav, K., Pavel, K., Jaromir, L., Vaclav, Z., & Ladislav, K., (2012). Identification of potential sources of thymoquinone and related compounds in Asteraceae, Cupressaceae, Lamiaceae, and Ranunculaceae families. *Central European Journal of Chemistry, 10*(6), 1899–1906.
8. Aftab, A., Asif, H., Mohd, M., Shah, A. K., Abul, K. N., Nasir, A. S., Zoheir, A. D., & Firoz, A., (2013). A review on therapeutic potential of *Nigella sativa*: A miracle herb. *Asian Pacific Journal of Tropical Biomedicine, 3*(5), 337–352.

9. Salem, M. L., (2005). Immunomodulatory and therapeutic properties of the *Nigella sativa* L. seed. *Int. Immunopharmacol., 5*, 1749–1770.
10. Khan, M. A., (1999). Chemical composition and medicinal properties of *Nigella sativa* Linn. *Inflammopharmacology, 7*, 15–35.
11. Gali-Muhtasib, H., El-Najjar, N., & Schneider-Stock, R., (2006). The medicinal potential of black seed (*Nigella sativa*) and its components. *Adv. Phytomed., 2*, 133–153.
12. Randolph, R. J. A., & Hajara H. A., (2018). Chemical properties of thymoquinone, a monoterpene isolated from the seeds of *Nigella sativa* Linn. *Pharmacological Research, 133*, 151.
13. Nagi, M. N., & Almakki, H. A., (2009). Thymoquinone supplementation induces quinone reductase and glutathione transferase in mice liver: Possible role in protection against chemical carcinogenesis and toxicity. *Phytother. Res., 23*, 1295–1298.
14. El-Dakhakhny, M., (1965). Studies on the Egyptian *Nigella sativa* L. Some pharmacological properties of the seeds' active principle in comparison to its dihydrocompound and its polymer. *Arzneimittelforschung, 15*, 1227–1229.
15. Mansour, M. A., Ginawi, O. T., El-Hadiyah, T., El-Khatib, A. S., Al-Shabanah, O. A., & Al-Sawaf, H. A., (2001). Effects of volatile oil constituents of *Nigella sativa* on carbon tetrachloride-induced hepatotoxicity in mice: Evidence for antioxidant effects of thymoquinone. *Res. Commun. Mol. Pathol. Pharmacol., 110*, 239–251.
16. Mansour, M. A., Nagi, M. N., El-Khatib, A. S., & Al-Bekairi, A. M., (2002). Effects of thymoquinone on antioxidant enzyme activities, lipid peroxidation and DT-diaphorase in different tissues of mice: A possible mechanism of action. *Cell Biochem. Funct., 20*, 143–151.
17. Kanter, M., (2008). *Nigella sativa* and derived thymoquinone prevents hippocampal neurodegeneration after chronic toluene exposure in rats. *Neurochem. Res., 33*, 579–588.
18. Al-Majed, A. A., Al-Omar, F. A., & Nagi, M. N., (2006). Neuroprotective effects of thymoquinone against transient forebrain ischemia in the rat hippocampus. *Eur. J. Pharmacol., 543*, 40–47.
19. El-Saleh, S. C., Al-Sagair, O. A., & Al-Khalaf, M. I., (2004). Thymoquinone *and Nigella sativa* oil protection against methionine-induced hyperhomocysteinemia in rats. *Int. J. Cardiol., 93*, 19–23.
20. Kanter, M., (2011). Thymoquinone attenuates lung injury induced by chronic toluene exposure in rats. *Toxicol. Ind. Health, 27*, 387–395.
21. Alkharfy, K. M., Ahmad, A., Khan, R. M., & Al-Shagha, W. M., (2015). Pharmacokinetic plasma behaviors of intravenous and oral bioavailability of thymoquinone in a rabbit model. *Eur. J. Drug Metab. Pharmacokinet, 40*(3), 319–323.
22. Tahereh, F., Saeed, S., & Abasalt, B., (2017). An overview on cardioprotective and anti-diabetic effects of thymoquinone. *Asian Pacific Journal of Tropical Medicine, 10*(9), 849–854.
23. Tahereh, F., Saeed, S., Soroush, H., & Azimi-Nezhad, M., (2017). Therapeutic effects of thymoquinone for the treatment of central nervous system tumors: A review. *Biomedicine and Pharmacotherapy, 96*, 1440–1444.
24. Nagi, M. N., & Mansour, M. A., (2000). Protective effect of thymoquinone against doxorubicin-induced cardiotoxicity in rats: A possible mechanism of protection. *Pharmacol. Res., 41*, 283–289.

25. Badary, O. A., Taha, R. A., Gamal El-Din, A. M., & Abdel-Wahab, M. H., (2003). Thymoquinone is a potent superoxide anion scavenger. *Drug Chem. Toxicol., 26,* 87–98.

26. Daba, M. H., & Abdel-Rahman, M. S., (1998). Hepatoprotective activity of thymoquinonein isolated rat hepatocytes. *Toxicol. Lett., 95,* 23–29.

27. Badary, O. A., & Gamal El-Din, A. M., (2001). Inhibitory effects of thymoquinone against 20-methylcholanthrene-induced fibrosarcoma tumorigenesis. *Cancer Detect. Prev., 25,* 362–368.

28. Kanter, M., Demir, H., Karakaya, C., & Ozbek, H., (2005). Gastro protective activity of *Nigella sativa* L oil and its constituent, thymoquinone against acute alcohol-induced gastric mucosal injury in rats. *World J. Gastroenterol., 11,* 6662–6666.

29. Elbarbry, F., Ragheb, A., Marfleet, T., & Shoker, A., (2012). Modulation of hepatic drug metabolizing enzymes by dietary doses of thymoquinone in female New Zealand White rabbits. *Phytother. Res., 26,* 1726–1730.

30. Houghton, P. J., Zarka, R., De las Heras, B., & Hoult, J. R., (1995). Fixed oil of *Nigella sativa* and derived thymoquinone inhibit eicosanoid generation in leukocytes and membrane lipid peroxidation. *Planta Med., 61,* 33–36.

31. Ammar, E. S. M., Gameil, N. M., Shawky, N. M., & Nader, M. A., (2011). Comparative evaluation of anti-inflammatory properties of thymoquinone and curcumin using an asthmatic murine model. *Int. Immunopharmacol., 11,* 2232–2236.

32. El Gazzar, M., El Mezayen, R., Marecki, J. C., Nicolls, M. R., Canastar, A., & Dre-skin, S. C., (2006). Anti-inflammatory effect of thymoquinone in a mouse model of allergic lung inflammation. *Int. Immunopharmacol., 6,* 1135–1142.

33. El Gazzar, M., El Mezayen, R., Nicolls, M. R., Marecki, J. C., & Dreskin, S. C., (2006). Down-regulation of leukotriene biosynthesis by thymoquinone attenuates airway inflammation in a mouse model of allergic asthma. *Biochim. Biophys. Acta, 1760,* 1088–1095.

34. El Mezayen, R., El Gazzar, M., Nicolls, M. R., Marecki, J. C., Dreskin, S. C., & Nomiyama, H., (2006). Effect of thymoquinone on cyclooxygenase expression and prostaglandin production in a mouse model of allergic airway inflammation. *Immunol. Lett., 106,* 72–81.

35. Hayat, K., Asim, M. B., Nawaz, M., Li, M., Zhang, L., & Sun, N., (2011). Ameliorative effect of thymoquinone on ovalbumin-induced allergic conjunctivitis in Balb/c mice. *Curr. Eye Res., 36,* 591–598.

36. Umar, S., Zargan, J., Umar, K., Ahmad, S., Katiyar, C. K., & Khan, H. A., (2012). Modulation of the oxidative stress and inflammatory cytokine response by thymoquinone in the collagen induced arthritis in Wistar rats. *Chem. Biol. Interact., 197,* 40–46.

37. Tekeoglu, I., Dogan, A., & Demiralp, L., (2006). Effects of thymoquinone (volatile oil of black cumin) on rheumatoid arthritis in rat models. *Phytother. Res., 20,* 869–871.

38. Vaillancourt, F., Silva, P., Shi, Q., Fahmi, H., Fernandes, J. C., & Benderdour, M., (2011). Elucidation of molecular mechanisms underlying the protective effects of thymoquinone against rheumatoid arthritis. *J. Cell. Biochem., 112,* 107–117.

39. Mohamed, A., Shoker, A., Bendjelloul, F., Mare, A., Alzrigh, M., Benghuzzi, H., et al., (2003). Improvement of experimental allergic encephalomyelitis (EAE) by thymoquinone: An oxidative stress inhibitor. *Biomed. Sci. Instrum., 39,* 440–445.

40. Mohamed, A., Afridi, D. M., Garani, O., & Tucci, M., (2005). Thymoquinone inhibits the activation of NF-kappaB in the brain and spinal cord of experimental autoimmune encephalomyelitis. *Biomed. Sci. Instrum., 41*, 388–393.

41. Racoma, I. O., Meisen, W. H., Wang, Q. E., Kaur, B., & Wani, A. A., (2013). Thymoquinone inhibits autophagy and induces cathepsin-mediated, caspase-independent cell death in glioblastoma cells. *PLoS One, 8*(9), e72882.

42. Ashour, A. E., Ahmed, A. F., Kumar, A., Zoheir, K. M., Aboul-Soud, M. A., Ahmad, S. F., Attia, S. M., et al., (2016). Thymoquinone inhibits growth of human medulloblastoma cells by inducing oxidative stress and caspase dependent apoptosis while suppressing NF-κB signaling and IL-8 expression. *Mol. Cell. Biochem., 416*(1–2), 141–155.

43. Alobaedi, O. H., Talib, W. H., & Basheti, I. A., (2017). Antitumor effect of thymoquinone combined with resveratrol on mice transplanted with breast cancer. *Asian Pac. J. Trop. Med., 10*(4), 400–408.

44. Rajput, S., Kumar, B. N., Dey, K. K., Pal, I., Parekh, A., & Mandal, M., (2013). Molecular targeting of Akt by thymoquinone promotes G(1) arrest through translation inhibition of cyclin D1 and induces apoptosis in breast cancer cells. *Life Sci., 93*(21), 783–790.

45. Yang, J., Kuang, X. R., Lv, P. T., & Yan, X. X., (2015). Thymoquinone inhibits proliferation and invasion of human non-small-cell lung cancer cells via ERK pathway. *Tumor Biol., 36*(1), 259–269.

46. Jafri, S. H., Glass, J., Shi, R., Zhang, S., Prince, M., & Kleiner-Hancock, H., (2010). Thymoquinone and cisplatin as a therapeutic combination in lung cancer: *In vitro* and *in vivo. J. Exp. Clin. Cancer Res., 29*, 87.

47. Lei, X., Lv, X., Liu, M., Yang, Z., Ji, M., Guo, X., & Dong, W., (2012). Thymoquinone inhibits growth and augments 5-fluorouracil-induced apoptosis in gastric cancer cells both *in vitro* and *in vivo. Biochem. Biophys. Res. Commun., 417*(2), 864–868.

48. Feng, L. M., Wang, X. F., & Huang, Q. X., (2017). Thymoquinone induces cytotoxicity and reprogramming of EMT in gastric cancer cells by targeting PI3K/Akt/mTOR pathway. *J. Biosci., 42*(4), 547–554.

49. Zhang, M., Du, H., Huang, Z., Zhang, P., Yue, Y., Wang, W., Liu, W., Zeng, J., Ma, J., Chen, G., Wang, X., & Fan, J., (2018). Thymoquinone induces apoptosis in bladder cancer cell via endoplasmic reticulum stress-dependent mitochondrial pathway. *Chem. Biol. Interact., 292*, 65–75.

50. Mu, H. Q., Yang, S., Wang, Y. J., & Chen, Y. H., (2012). Role of NF-kappaB in the anti-tumor effect of thymoquinone on bladder cancer. *Zhonghua Yi Xue Za Zhi., 92*(6), 392–396.

51. Norwood, A. A., Tan, M., May, M., Tucci, M., & Benghuzzi, H., (2006). Comparison of potential chemotherapeutic agents, 5-fluoruracil, green tea, and thymoquinone on colon cancer cells. *Biomed. Sci. Instrum., 42*, 350–356.

52. Zhang, L., Bai, Y., & Yang, Y., (2016). Thymoquinone chemosensitizes colon cancer cells through inhibition of NF-κB. *Oncol. Lett., 12*(4), 2840–2845.

53. Kou, B., Liu, W., Zhao, W., Duan, P., Yang, Y., Yi, Q., Guo, F., Li, J., Zhou, J., & Kou, Q., (2017). Thymoquinone inhibits epithelial-mesenchymal transition in prostate cancer cells by negatively regulating the TGF-beta/Smad2/3 signaling pathway. *Oncol. Rep., 38*(6), 3592–3598.

54. Roepke, M., Diestel, A., Bajbouj, K., Walluscheck, D., Schonfeld, P., Roessner, A., Schneider-Stock, R., & Gali-Muhtasib, H., (2007). Lack of p53 augments thymoquinone induced apoptosis and caspase activation in human osteosarcoma cells. *Cancer Biol. Ther.*, *6*(2), 160–169.

55. Yasmina, K. M., & Heba, M. A. A., (2019). Cancer: Thymoquinone antioxidant/pro-oxidant effect as potential anticancer remedy. *Biomedicine and Pharmacotherapy*, *115*(108783), 1–14.

56. Juthika, K., Kyung-Soo, C., Okezie, I. A., & Joydeb, K. K., (2014). Mechanistic perspectives on cancer chemoprevention/chemotherapeutic effects of thymoquinone. *Mutation Research*, *768*, 22–34.

57. Gurung, R. L., Lim, S. N., Khaw, A. K., Soon, J. F., Shenoy, K., Mohamed, S. A., Jayapal, M., Sethu, S., Baskar, R., & Hande, M. P., (2010). Thymoquinone induces telomere shortening, DNA damage, and apoptosis in human glioblastoma cells. *PLoS One*, *5*(8), e12124.

58. Talib, W. H., (2017). Regressions of breast carcinoma syngraft following treatment with piperine in combination with thymoquinone. *Sci. Environ.*, *3*, 85.

59. Ganji-Harsini, S., Khazaei, M., Rashidi, Z., & Ghanbari, A., (2016). Thymoquinone could increase the efficacy of Tamoxifen induced apoptosis in human breast cancer cells: An *in vitro* study. *Cell J.*, *18*(2), 245–254.

60. Şakalar, C., İzgi, K., İskender, B., Sezen, S., Aksu, H., Çakır, M., Kurt, B., Turan, A., & Canatan, H., (2016). The combination of thymoquinone and paclitaxel shows anti-tumor activity through the interplay with apoptosis network in triple-negative breast cancer. *Tumor Biol.*, *37*(4), 4467–4477.

61. Dehghani, H., Hashemi, M., Entezari, M., & Mohsenifar, A., (2015). The comparison of anticancer activity of thymoquinone and nanothymoquinone on human breast adenocarcinoma, *Iran J. Pharm. Res.*, *14*(2), 539–546.

62. Muhammad, I., Abdur, R., Imtiaz, A. K., Muhammad, S., Tahira, B. Q., Sri, F., Abu-Izneid, T., Ali, I., Khaliq, U. R., & Tanweer, A. G., (2018). Thymoquinone: A novel strategy to combat cancer: A review. *Biomedicine and Pharmacotherapy*, *106*, 390–402.

63. Peng, L., Liu, A., Shen, Y., Xu, H. Z., Yang, S. Z., Ying, X. Z., et al., (2013). Antitumor and anti-angiogenesis effects of thymoquinone on osteosarcoma through the NF-κB pathway. *Oncol. Rep.*, *29*(2), 571–578.

64. Ng, W. K., Saiful-Yazan, L., Yap, L. H., WanNor-Hafiza, W. A., How, C. W., & Abdullah, R., (2015). Thymoquinone-loaded nanostructured lipid carrier exhibited cytotoxicity towards breast cancer cell lines (MDA-MB-231 and MCF-7) and cervical cancer cell lines (HeLa and SiHa). *Biomed. Res. Int.*, p. 10.

65. Sakalar, C., Yuruk, M., Kaya, T., Aytekin, M., Kuk, S., & Canatan, H., (2013). Pronounced transcriptional regulation of apoptotic and TNF-NF-kappa-B signaling genes during the course of thymoquinone mediated apoptosis in HeLa cells. *Mol. Cell Biochem.*, *383*(1–2), 243–251.

66. Hasan, T. N., Shafi, G., Syed, N. A., Alfawaz, M. A., Alsaif, M. A., Munshi, A., Lei, K. Y., & Alshatwi, A. A., (2013). Methanolic extract of Nigella sativa seed inhibits SiHa human cervical cancer cell proliferation through apoptosis. *Nat. Prod. Commun.*, *8*(2), 213–216.

67. Ulasli, S. S., Celik, S., Gunay, E., Ozdemir, M., Hazman, O., Ozyurek, A., Koyuncu, T., & Unlu, M., (2013). Anticancer effects of thymoquinone, caffeic acid phenethyl ester and resveratrol on A549 non-small cell lung cancer cells exposed to benzo(a) pyrene. *Asian Pac, J. Cancer Prev., 14*(10), 6159–6164.

68. Mohamed, A. M., Refaat, B. A., El-Shemi, A. G., Kensara, O. A., Ahmad, J., & Idris, S., (2017). Thymoquinone potentiates chemoprotective effect of vitamin D3 against colon cancer: A pre-clinical finding. *Am. J. Transl. Res., 9*(2), 774–790.

69. Hsu, H. H., Chen, M. C., Day, C. H., Lin, Y. M., Li, S. Y., Tu, C. C., Padma, V. V., Shih, H. N., Kuo, W. W., & Huang, C. Y., (2017). Thymoquinone suppresses migration of LoVo human colon cancer cells by reducing prostaglandin E2 induced COX-2 activation. *World J. Gastroenterol., 23*(7), 1171–1179.

70. Frohlich, T., Ndreshkjana, B., Muenzner, J. K., Reiter, C., Hofmeister, E., Mederer, S., Fatfat, M., El-Baba, C., Gali-Muhtasib, H., Schneider-Stock, R., & Tsogoeva, S. B., (2017). Synthesis of novel hybrids of thymoquinone and artemisinin with High activity and selectivity against colon cancer. *Chem. Med. Chem., 12*(3), 226–234.

71. Ojha, S., Azimullah, S., Mohanraj, R., Sharma, C., Yasin, J., Arya, D. S., et al., (2015). Thymoquinone protects against myocardial ischemic injury by mitigating oxidative stress and inflammation. *Evid. Based Complement. Alternat. Med.,* p. 143629.

72. Nagi, M. N., AI-Shabanah, O. A., Hafez, M. M., & Sayed-Ahmed, M. M., (2011). Thymoquinone supplementation attenuates cyclophosphamide induced cardiotoxicity in rats. *J. Biochem. Mol. Toxicol., 25*(3), 135–142.

73. Al-Shabanah, O. A., Badmy, O. A., Nagi, M. N., Ai-Gharably, N. M., Ai-Rikabi, A. C., & Al-Bekairi, A. M., (1998). Thymoquinone protects against doxorubicin-induced cardiotoxicity without compromising its antitumor activity. *J. Exp. Clin. Cancer Res., 17*(2), 193–198.

74. Nader, M. A., El-Agamy, D. S., & Suddek, G. M., (2010). Protective effects of propolis and thymoquinone on development of atherosclerosis in cholesterolfed rabbits. *Arch. Pharm. Res., 33*, 637–643.

75. Ragheb, A., Elbarbry, F., Prasad, K., Mohamed, A., Ahmed, M. S., & Shoker, A., (2008). Attenuation of the development of hypercholesterolemic atherosclerosis by thymoquinone. *Int. J. Angiol., 17*(4), 186–192.

76. Ragheb, A., Attia, A., Elbarbry, F., Prasad, K., & Shoker, A., (2011). Attenuated combined action of cyclosporine A and hyperlipidemia on atherogenesis in rabbits by thymoquinone. *Evid. Based Complement Altern. Med.,* p. 620319.

77. Ahmad, S., & Beg, Z. H., (2013). Hypolipidemic and antioxidant activities of thymoquinone and limonene in atherogenic suspension fed rats. *Food Chem., 138*(2–3), 1116–1124.

78. Umeno, A., Horie, M., Murotomi, K., Nakajima, Y., & Yoshida, Y., (2016). Antioxidative and antidiabetic effects of natural polyphenols and isoflavones. *Molecules, 21*(6), E708.

79. AI-Trad, B., AI-Batayneh, K., EI-Metwally, S., Alhazimi, A., Ginawi, I., Alaraj, M., et al., (2016). *Nigella sativa* oil and thymoquinone ameliorate albuminuria and renal extracellular matrix accumulation in the experimental diabetic rats. *Eur. Rev. Med. Phamacol. Sci., 20*(12), 2680–2688.

80. Pei, X., Li, X., Chen, H., Han, Y., & Fan, Y., (2016). Thymoquinone inhibits angiotensin II-Induced proliferation and migration of vascular smooth muscle cells through the AMPK/PPARg/PGC-1a pathway. *DNA Cell Biol., 35*(8), 426–433.

81. Fouad, A. A., & Alwadani, F., (2015). Ameliorative effects of thymoquinone against eye lens changes in streptozotocin diabetic rats. *Environ. Toxicol. Pharmacol., 40*(3), 960–965.

82. Al-Wafai, R. J., (2013). Nigella sativa and thymoquinone suppress cyclooxygenase-2 and oxidative stress in pancreatic tissue of streptozotocin-induced diabetic rats. *Pancreas, 42*(5), 841–849.

83. Fararh, K. M., Shimizu, Y., Shiina, T., Nikami, H., Ghanem, M. M., & Takewaki, T., (2005). Thymoquinone reduces hepatic glucose production in diabetic hamsters. *Res. Vet. Sci., 79*(3), 219–223.

84. Badr, G., Mahmoud, M. H., Farhat, K., Waly, H., AI-Abdin, O. Z., & Rabah, O. M., (2013). Maternal supplementation of diabetic mice with thymoquinone protects their offspring from abnormal obesity and diabetes by modulating their lipid profile and free radical production and restoring lymphocyte proliferation via PI3K/AKT signaling. *Lipids Health Dis., 12*, 37.

85. Sankaranarayanan, C., & Pari, L., (2011). Thymoquinone ameliorates chemical induced oxidative stress and β-cell damage in experimental hyperglycemic rats. *Chem. Biol. Interact., 190*(2–3), 148–154.

86. Abdelmeguid, N. E., Fakhomy, R., Kamal, S. M., & AI-Wafai, R. J., (2010). Effects of *Nigella sativa* and thymoquinone on biochemical and subcellular changes in pancreatic b-cells of streptozotocin-induced diabetic rats. *J. Diabetes, 2*(4), 256–266.

87. Kouidhi, B., Zmantar, T., Jrah, H., Souiden, Y., Chaieb, K., Mahdouani, K., et al., (2011). Antibacterial and resistance-modifying activities of thymoquinone against oral pathogens. *Ann. Clin. Microbiol. Antimicrob., 10*, 29.

88. Harzallah, H. J., Kouidhi, B., Flamini, G., Bakhrouf, A., & Mahjoub, T., (2011). Chemical composition, antimicrobial potential against cariogenic bacteria and cytotoxic activity of Tunisian *Nigella sativa* essential oil and thymoquinone. *Food Chem., 129*, 1469–1474.

89. Chaieb, K., Kouidhi, B., Jrah, H., Mahdouani, K., & Bakhrouf, A., (2011). Antibacterial activity of Thymoquinone, an active principle of *Nigella sativa* and its potency to prevent bacterial biofilm formation. *BMC Complement. Altern. Med., 11*, 29.

90. Halawani, E., (2009). Antibacterial activity of thymoquinone and thymohydroquinone of *Nigella sativa* L. and their interaction with some antibiotics. *Adv. Biol. Res., 3*, 148–152.

91. Aljabre, S. H. M., Randhawa, M. A., Akhtar, N., Alakloby, O. M., Alqurashi, A. M., & Aldossary, A., (2005). Antidermatophyte activity of ether extract of *Nigella sativa* and its active principle, thymoquinone. *J. Ethnopharmacol., 101*, 116–119.

92. Sheikh, B. Y., Taha, M. M. E., Koko, W. S., & Abdelwahab, S. I., (2015). Antimicrobial effects of thymoquinone on *Entamoeba histolytica* and *Giardia lamblia*. *Pharm. J., 8*, 168–170.

93. Umar, S., Shah, M., Munir, M., Yaqoob, M., Fiaz, M., Anjum, S., et al., (2016). Synergistic effects of thymoquinone and curcumin on immune response and anti-viral activity against avian influenza virus (H9N2) in turkeys. *Poult. Sci., 95*, 1513–1520.

Index

EMPOWERING ARTIFICIAL INTELLIGENCE THROUGH MACHINE LEARNING

New Advances and Applications